Troubleshooting and Repairing Electronic Circuits
2nd Edition

Robert L. Goodman

TAB BOOKS Inc.
Blue Ridge Summit, PA

SECOND EDITION
FIRST PRINTING

Copyright © 1990 by TAB BOOKS Inc.
Printed in the United States of America

Library of Congress Cataloging in Publication Data

Goodman, Robert L.
Troubleshooting and repairing electronic circuits / by Robert L.
Goodman.

p. cm.
Rev. ed. of: How to troubleshoot & repair electronic circuits. 1st
ed. 1981.

ISBN 0-8306-9258-4 ISBN 0-8306-3258-1 (pbk.)
1. Electronic apparatus and appliances—Maintenance and repair.
I. Goodman, Robert L. How to troubleshoot & repair electronic
circuits. II. Title.

TK7870.2.G665   1989   89-39628
621.381′5′0288—dc20   CIP

TAB BOOKS Inc. offers software for
sale. For information and a catalog,
please contact TAB Software Department,
Blue Ridge Summit, PA 17294-0850.

Questions regarding the content of this book
should be addressed to:

Reader Inquiry Branch
TAB BOOKS Inc.
Blue Ridge Summit, PA 17294-0214

Acquisitions Editor: Roland S. Phelps
Production: Katherine Brown
Cover Design: Lori E. Schlosser

Contents

Acknowledgments

I wish to thank the following companies and personnel.

Harris Corp.
Magnavox Corp.
Radio Shack, Dave Gunzel
Sencore, Inc., Mr. Bowden and Greg Carey
Quasar Electronics, Charlie Howard
Sony Corporation of America, Howard L. Katz
Hewlett-Packard
B&K Dynascan Corp., Myren E. Bond
Zenith Radio Corp., James F. White

Introduction

In this revised book you will find selected electronic circuits, ways to troubleshoot them, and known component failures. Also, some actual case-history problems are given. In some TV circuits a photo of the picture symptoms is provided. Voltage, resistance, and oscilloscope waveforms are included where applicable.

Included in this book you will find the tube type, transistor, IC, microprocessor, analog, and digital logic circuits for various consumer electronic devices. Old and new circuits are described along with the types of electronic equipment the circuits are found in.

In this second edition, you will find expanded information on newer electronic circuits in each chapter. This includes electronic TV tuners, state-of-the-art TV remote control systems, digital TV circuits, TV stereo sound, compact disc (CD) player circuits, super VCR circuits, troubleshooting microprocessors with signature analysis, and much more.

The electronic circuit reference data in this book provides you with up-to-date troubleshooting and theory-of-operation information to assist you in keeping on top of today's advanced space-age electronic systems.

In all, 40 years of electronic troubleshooting experience and know-how has been collected and presented in this reference book for your convenience and enlightenment.

1

Color TV Tuners

In this chapter we will cover operation and troubleshooting information on various electronic tuners used in assorted makes of color TV receivers. The circuits will include the VHF and UHF tuners along with their control center circuits. Some will be the manual channel selector control units, while others will be remote controlled via all electronic switching IC and transistor devices. Various dc voltage control and AFC (automatic frequency control) circuit schemes for electronic tuners will be reviewed.

THE SYLVANIA MANUAL ELECTRONIC TUNING SYSTEM

Some Sylvania color sets have conventional two knob UHF and VHF tuning systems, but others have a combination UHF/VHF varactor tuner with a one knob addressing system. One channel selector and one fine tuning knob are used to change and adjust both UHF and VHF stations.

The tuning voltage applied to the UHF/VHF tuner is varied in two ways. Note the tuner control circuit shown in Fig. 1-1. Band switches apply the 24 volt B+ line to either the UHF or VHF portions of the tuner. The variable tuning voltage comes from the 112 volt B+ line and is dropped and regulated by a 33 volt zener on the tuner cluster assembly. This 33 volts is varied by adjusting variable resistors mounted in a tuning assembly that looks like a small barrel type tuner. Instead of many contacts on each of the strips, as with a conventional barrel type tuner, these strips have only three contacts on the variable 10 k ohm resistor. The voltage output from the variable resistors, ranges from +1 volt to +28 volts. As the channel selector is rotated, a different dial appears through the viewing window. This dial indicates the channel ranges in either the UHF or VHF spectrums. As the fine tuning knob is varied, a red pointer moves up and

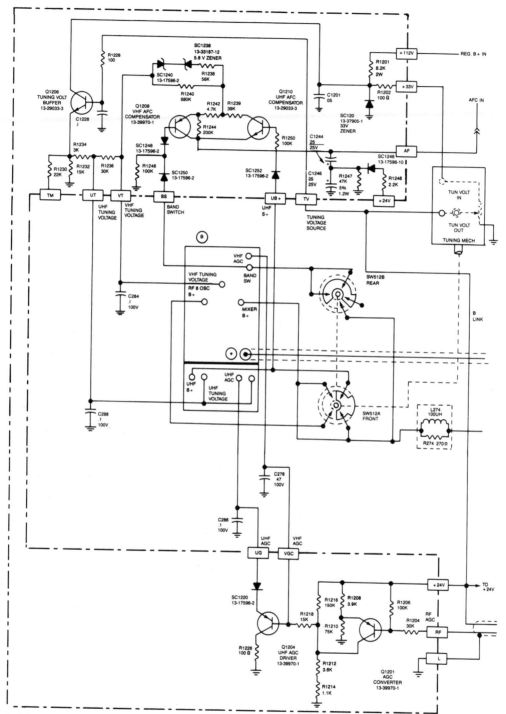

Fig. 1-1. Sylvania manual control electronic tuner system.

down the indicator dial showing the channel being tuned. This dial is located below the channel knob. A second channel indicator window is visible above the channel knob. It indicates all of the VHF channels and has positions to insert the UHF channel numbers that the set will be tuned to in the viewing area. Once the channels have been programmed, the viewer turns the channel selector until the desired station number appears in the top viewing window.

In addition to varying the tuning voltage with the variable resistors, the tuning voltage is also varied slightly by the AFC system. The AFC voltage is applied to a printed circuit board mounted on the tuner cluster that has the various circuits required to drive the varactor tuner. The AFC voltage is fed to Q1208 the VHF compensator and Q1210 the UHF AFC compensator. The output of these two transistors varies the tuning voltage slightly to produce automatic fine tuning action.

The output of the varied 33 volts from the addressing turret is applied to Q1206 the tuning voltage buffer transistor. The output of this emitter follower transistor is varied slightly by the AFC system and then applied to the varactor tuner itself.

The RF AGC voltage from the set is applied to Q1201 the AGC converter transistor and Q1204 the UHF AGC driver transistor, both of which are mounted on the PC board on the tuner cluster. The RF AGC voltage outputs of these two transistors are fed to the UHF and VHF sections of the varactor tuner.

The AGC voltages measured with a signal (the set tuned to a local station) and without a signal (the set tuned to an unused channel) are RF AGC input +8 volts with signal, +4 volts without signal. VHF tuner input +4 volts with signal, +8 volts without. UHF tuner input +4 volts with a signal and +9 volts without a signal.

ZENITH DIRECT ACCESS TUNING SYSTEM

Looking at the block diagram in Fig. 1-2, you will see seven major block sections that are required to make the direct access tuning system operational. As you may guess, the heart of the system is the new microprocessor module. This A-7892 module consists of six integrated circuits and 13 transistors with all of the associated circuitry. The 175-5104 direct access tuning unit package consists of the new microprocessor module (A-7892), keyboard channel selector, prescaler, and the space command remote receiver. There is also a manual version of this direct access tuning system.

Channel Selection Operation

The receiver-mounted keyboard consists of 15 snap-action push buttons. This type of push button switch offers a positive user feel and has the following features:

- Separate on/off buttons (button must be held a moment for on/off action).
- 0 to 9 channel selector buttons.
- Separate volume up/down buttons.
- Enter button.

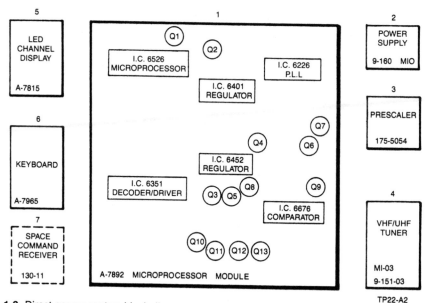

Fig. 1-2. Direct access system block diagram.

A channel can be selected by pushing the desired channel number and then pressing the *enter* key. Also, a 0 does not have to be entered for a single channel number. If a channel number is selected without pressing the enter key, the display will show the new number, but will return to the original number after 4.5 seconds have elapsed.

This direct access system will also tune the cable TV channels that are referred to as the super band channels J through W and the mid-channels A through I. The mid- and super channels are tuned on channels 14 to 36.

Direct Access Functional Blocks

As we know, the microprocessor IC is the heart of this complete tuning system. It is essentially a small computer which controls all of the functions of the tuning system.

- It constantly scans the receiver mounted keyboard looking for a contact closure.
- It constantly monitors the remote receiver output looking for a remote key closure.
- It controls the operation of the LED channel display through a decoder/driver IC.
- It provides B+ switching and bandswitching for the VHF and UHF tuners.
- It controls the operation of a phase-locked-loop IC which results in precise control of the tuner oscillator frequency. See Fig. 1-3.

This tuning system is referred to as a frequency synthesis tuning system. In a voltage synthesis type, a precise, stable tuning voltage is generated and used to control the tuner oscillator. This type of system depends heavily on conventional AFC operation for oscillator frequency stability.

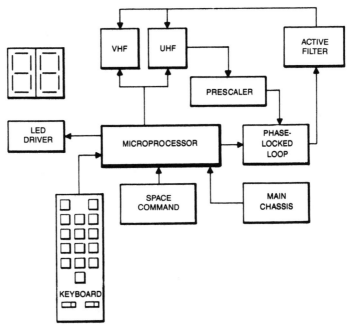

Fig. 1-3. Functional block diagram of direct access system.

In a frequency synthesis system, a precise frequency is generated and maintained. When a channel is selected, the microprocessor knows what the correct FCC designated frequency should be for that channel. It develops a tuning voltage that will put the oscillator at the correct frequency. It constantly monitors the oscillator frequency by dividing, counting and comparing it to a reference, which results in automatic correction of the tuning voltage, assuring oscillator stability.

As you will note in Fig. 1-4, a crystal-controlled oscillator, operating at 3.581055 MHz, is divided by a 14 stage reference divider in the phase-locked-loop IC, to provide an output frequency of 976.5625 Hz. This frequency, rounded off to 1 kHz, becomes the reference base which is applied to one input of a comparator.

The VHF or UHF oscillator frequencies are applied to a *prescaler* circuit. This prescaler IC and its associated circuitry, is required because the normal VHF and UHF oscillator frequencies for each channel are much too high to be used as a comparison in the phase-locked-loop IC. The prescaler drives the oscillator signal down by 256.

Referring to Fig. 1-4 again, we see the divide-down oscillator signal is applied to a programmable divider in the phase-locked-loop IC. Here, it is further divided by a variable ratio, to produce an output frequency which becomes the second input to the comparator. The comparator produces an output which is dependent on the differences between the crystal controlled reference signal and the divided down oscillator signal.

The comparator output is a series of pulses with variable duty cycles. These pulses are fed to an active filter circuit. The active filter smoothes these pulses and produces a dc output with a negligible amount of 1 kHz ripple. This dc voltage is in fact the tuning

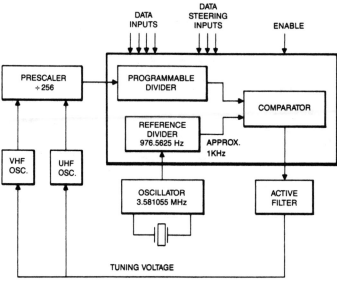

Fig. 1-4. Phase-locked-loop block diagram.

voltage which is applied to the varactor diodes in both tuners. In this manner, the tuner oscillator assumes the stability of the crystal reference.

In order for the comparator to operate correctly, the two input frequencies (tuner oscillator and reference) should be about equal (1 kHz). Since the oscillator frequencies are different for each channel, a different divide ratio must be used for each channel. For example, the channel 2 local oscillator frequency is 101 MHz. In order to yield a 1 kHz output, the signal must be divided by 101,000. The channel 13 local oscillator is at 257 MHz. This frequency must be divided by 257,000 to produce the same 1 kHz signal. This task is accomplished in the programmable divider.

The divide ratio is determined by information provided by the microprocessor via four data input control lines. Each time a channel is selected, the microprocessor will put the correct binary control signal on each of the four input lines to establish the correct divide ratio for that particular channel.

Three data steering input lines are used to determine to which internal registers the binary signals belong. The microprocessor performs many other functions besides controlling the phase-locked-loop IC. Because of this, there may be information on the control input lines which is not meant for the phase-locked-loop IC. The eighth input from the microprocessor is the *enable*. Information on this line permits the phase-locked-loop IC to receive the correct data and exclude all other information.

The 1 kHz pulse width modulated output from the phase-locked-loop IC is applied to the base of Q1. A Darlington stage is used here to maintain an extremely high input impedance. The output of the collector of Q2 is fed back to Q1 via C14, C15 and R67. This filtered DC voltage at the collector is then routed through Q12, an emitter follower, to provide a low impedance drive to the tuner.

Connected to the base of Q12 is an adjustable clamp circuit. This circuit does not allow the tuning voltage to go below 2.25 volts dc on the high VHF or mid-band CATV channels. Misadjustment of this clamp circuit can cause the tuner oscillator to stall, ultimately making the system to lock out.

The clamp voltage is set by a potentiometer located on the A-7892 microprocessor module.

The correct adjustment procedure is as follows:

- Set the mode switches to normal.
- Select a high (7 to 13) channel.
- Remove the shielded cable between the VHF tuner and prescaler.
- Adjust the potentiometer for 2.25 volts dc on the tuning voltage line.
- Reconnect the shielded cable.

Keyboard Scanning

The microprocessor constantly scans the keyboard by sending pulses on three output scan lines. See Fig. 1-5. These lines are arranged in such a way as to represent the vertical, or x axis of the keyboard (three vertical rows of keys). The horizontal, or y axis of the keyboard is represented by five output lines. When a key is depressed, the appropriate x-y contact is made. The contact closure transfers the pulse back to the microprocessor via one of the five output lines. The microprocessor detects the pulse, producing the desired digit on the display.

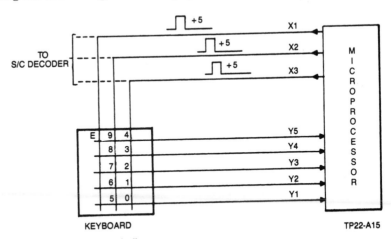

Fig. 1-5. Keyboard scanning block diagram.

Microprocessor Organization

All of these circuits are controlled by a microprocessor which is essentially a small computer. The internal organization of the microprocessor is shown in Fig. 1-6. The microprocessor consists of four major sections:

7

- The arithmetic logic unit (ALU) performs a sequence of operations determined by commands which are stored in various memory locations. It can do additions, subtractions, comparisons and so forth. It performs these operations one at a time.
- The accumulator (ACC) section of the IC is used by the (ALU) to temporarily store information while it performs its various duties.
- Read only memory (ROM) section contains all the instructions and sequencing information for the *arithmetic logic unit*. This program is designed to do a specific job and it enables a general purpose microprocessor to become suited for use in a tuning system. The ROM section of this microprocessor cannot be changed except by changing the metal mask used in fabricating the IC.
- The random access memory (RAM) however, can be changed. It is used to temporarily store information that can be altered as required. This is necessary because the microprocessor performs numerous functions on a time sharing basis. The microprocessor also has its own internal clock which is set by R6528. The nominal clock frequency is 375 kHz and can be measured at pin 28 of IC6526 with a low capacity probe.

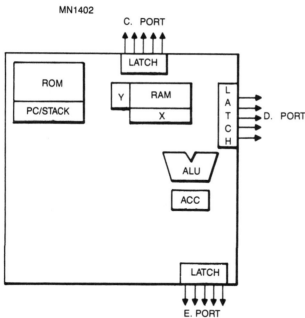

Fig. 1-6. Microprocessor organization diagram.

Microprocessor Voltage Information

All the integrated circuits in this system are powered from two 5 volt regulators on the microprocessor module. One regulator supplies all the LED current and is turned on and off with the TV set. The other regulator is powered from the remote control power supply which enables the system to have last channel memory.

NOTE OF CAUTION: The IC's in this system are mostly MOS devices and can be damaged by a static discharge. Use care when troubleshooting!

ELECTRONIC TUNING SYSTEM (RCA CTC-93 CHASSIS)

This RCA electronic tuner system utilizes a phase-locked loop (PLL) in conjunction with a digital frequency programmer to generate a number of discrete frequencies. To better understand the total system let's look at the basic PLL operation.

BASIC PHASE-LOCKED-LOOP THEORY

Phase-locked loops have been used for many electronic systems in the past few years. Phase locking is actually a technique of forcing the phase of an oscillator signal to exactly follow the phase of a reference signal. The PLL automatically locks onto and tracks a signal, even though its frequency changes. The PLL does all this with the help of its phase comparator and a voltage-controlled oscillator (VCO). The phase comparator samples the frequency of an input signal with that of a reference oscillator and produces an error voltage directly proportional to the difference between the frequencies of the two.

The error voltage serves two purposes. It is fed back to the VCO and changes its frequency to match that of the input signal. This feedback enables the PLL to lock onto and track the signal. The error voltage can also be considered a demodulated FM output since it varies directly with a shift in the input signal frequency. Thus, the error voltage from the phase comparator permits the PLL to lock onto a frequency, and to track it continually over a given range.

Basic Phase-Locked-Loop Operation

Lets now look at a digital phase-locked-loop or *frequency synthesis* (FS) phase-locked-loop operation. See Fig. 1-7.

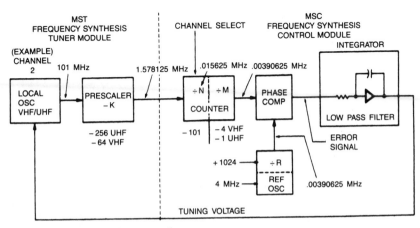

Fig. 1-7. Basic phase-locked-loop operation.

The FS tuning system for a TV receiver includes a phase-locked loop for synthesizing local VHF/UHF oscillator signals. When the RF input receives standard TV frequency carriers, the mixer combines them with local oscillator signals to form i-f signals having a picture carrier equal to the nominal i-f picture carrier frequency, 45.75 MHz. The PLL includes a local oscillator, a divide-by-K prescaler, a divide-by-N unit, and a divide-by-M unit, operating in conjunction with comparator, reference oscillator reference divider, and a low-pass filter.

The local oscillator, located in the MST tuner module (a voltage controlled type), generates a signal whose frequency is determined by the dc voltage applied to it from the low-pass filter located in the MSC tuner control module.

The output of the local oscillator is coupled to the divide-by-K prescaler which divides the frequency of the relatively high frequency local oscillator. This step is necessary to produce signals whose frequency is compatible with the operating frequency range of the portions of the tuning system following it. Factor K equals 256 for UHF and 64 for the VHF range.

The output of the divide-by-K prescaler is coupled to the divide-by-N unit. The divide-by-N unit performs a number of functions: One of these functions is to divide the frequency of the output signal of the K prescaler, by a number (N), which is equal to the frequency necessary to derive (in the closed loop system) the correct local oscillator signal for the desired TV channel.

The factor N is controlled in accordance with the channel selected via a channel selection unit. The channel select unit may be a keyboard type which can sequentially select the two decimal digits for the desired channel. The channel selection unit converts the selected two-digit decimal number into binary signals arranged in a binary coded decimal (BCD) format. The binary signals are partitioned into a group of four bits (binary digits) for the most significant digit (MSD) and another group of four bits for the least significant digit (LSD). The binary signals are coupled to the divide-by-N unit and are also coupled to a display unit which functions to provide an indication of the channel number selected.

A band decoder (part of the divide-by-N unit) determines the frequency band in which the selected channel resides. A band identification signal, indicating that the selected channel is in the VHF or UHF range, is coupled to the divide-by-K prescaler from the band switch stage to control the factor K (64 or 256).

The output of the divide-by-N unit is coupled to a divide-by-M stage which divides the frequency of the output signal of the divide-by-N unit by 1 for UHF channels and by 4 for VHF channels, in accordance with the state of the signal coupled to it from the band decoder. The output of the divide-by-M is coupled to the phase comparator which provides an output signal comprised of a series of pulses whose polarity and duty cycle represent the phase and/or frequency deviation between the output signal of the divide-by-M divider and the output signal of the reference divider R.

The output of the phase comparator is coupled to an active low-pass filter. This integrates the output signal of the phase comparator, to form a dc signal, which controls the frequency of the local oscillator.

The loop just described is arranged so that the low-pass filter couples a dc tuner control voltage to the local oscillator, which tends to minimize the frequency and phase differences between the output signals of the divide-by-M and reference divider.

AUTOMATIC OFFSET TUNING

The offset signals sometimes encountered with CATV converters, MATV systems, home video games, and video recorders fall in a range of ± 2 MHz around the FCC assigned VHF broadcast channel frequencies. These odd signals, however, must be tuned by the receiver.

The following details give the sequence of events involved with frequency synthesis automatic offset tuning. First, the desired VHF channel is selected. The synthesis system (PLL) tunes the FCC assigned frequency for that channel. When a phase lock has been completed, the system switches to an AFT mode for signal searching within the AFT pull-in range. At this time, if a station carrier frequency is within AFT range, the system stays in AFT mode and remains on channel. If a station carrier is not detected, with the oscillator tuned to this center synthesis frequency to produce 45.75 MHz, the synthesis system is again activated. Now, however, the system automatically offsets the local oscillator *plus* 1 MHz above the nominal FCC frequency and the AFT search mode is reactivated. If lock occurs, the system stabilizes in AFT mode, holding the offset carrier frequency. If lock is not achieved, the system will again go to the synthesis mode, this time tuning the local oscillator to a frequency 1 MHz below the FCC nominal oscillator frequency for the channel. The AFT search is again activated, seeking a station carrier within ± 1.25 MHz of this new synthesized oscillator frequency. If a carrier is available, AFT lock is established, otherwise, synthesis again retunes the oscillator to its nominal FCC channel frequency. The system will continue to perform alternate cycles of synthesis (PLL) and AFT in search of a station signal in the sequence just described.

The choice of ± 1.25 MHz for the AFT control limit prevents the possibility of AFT lockout to the lower adjacent sound carrier (-1.5 MHz) during dropout of the desired stations signal.

In summation:

- A channel is selected, synthesis occurs, tuning the local oscillator to the FCC nominal frequency to receive that channel.
- The AFT search mode is activated. If a station signal carrier is within ± 1.25 MHz, AFT lock-up is established and the system stabilizes in AFT mode.
- If during AFT search no carrier is available, synthesis is reengaged shifting the local oscillator 1 MHz high; AFT search is activated, seeking a station carrier. If available, AFT lock-up is established and the system stabilizes.
- If lock-up is not achieved, the system now tunes the oscillator 1 MHz lower than nominal. AFT search is set in the same sequence as in the previous high offset mode.

- A third failure to get an AFT lock will return the system back to the FCC nominal frequency. There it will do alternate cycles of synthesis (PLL) and AFT in search of a signal, but with no further stepping. If AFT lock is established for 3 seconds, further stepping is prevented.

DC TUNING VOLTAGE AND PRESCALER FREQUENCY

A dc tuning voltage and prescaler chart for VHF low, VHF high, and UHF is shown in Fig. 1-8. Use this voltage chart when troubleshooting electronic tuner troubles in this system. The dc tuning voltage is connected to the MST 001 tuner module at point P1. Tuning voltage is developed on the MSC control module, which is part of the synthesis tuning loop.

If a tuning voltage problem exists, remove power then check with an ohmmeter. Check point J1 to see if the tuning voltage is being shorted out. If not, sub in a battery or power supply. This will serve as a tuning voltage source and will check operation of the tuner. Also, check for correct band switch voltage. Now the desired station can be tuned in by adjusting the battery or power supply to the desired tuning voltage as shown in Fig. 1-8. With this setup the prescaler frequency can now be checked with a frequency counter. If the tuner module checks out good, the problem is probably in the MSC control module.

	CHANNEL	TUNING VOLTAGE (TYPICAL)	PRESCALER FREQ (MHZ)
V H F L O W	2	1.8	1.58
	3	3.3	1.67
	4	5.4	1.76
	5	11.4	1.92
	6	16.0	2.01
✱ ✱	✱✱✱✱✱✱✱✱✱	✱✱✱✱✱✱✱✱✱✱	✱✱✱✱✱✱✱✱✱✱
V H F H I G H	7	8.5	3.45
	8	9.6	3.54
	9	11.0	3.65
	10	12.6	3.73
	11	14.6	3.84
	12	17.5	3.92
	13	22.0	4.01
✱ ✱	✱✱✱✱✱✱✱✱✱	✱✱✱✱✱✱✱✱✱✱	✱✱✱✱✱✱✱✱✱✱
	14	1.8	2.02
	24	3.8	2.25
U H F	34	6.3	2.48
	43	8.4	2.69
	53	10.5	2.93
	63	13.4	3.16
	73	17.0	3.40
	83	24.0	3.63

Fig. 1-8. Tuning voltage and prescaler frequency chart.

Band Switching

Because of the limited tuning range of varactor diodes, it remains necessary to provide band-switching capability to select the low band VHF channels (2 through 6), the high band VHF channels (7 through 13) and the UHF channels (14 through 83). This is accomplished by applying +19 volts to the appropriate band-switching terminal. Refer to the band-switching chart in Fig. 1-9. Also included on the chart is the Mixer B+ voltage for each tuning band.

TV TUNING BAND	MIXER B+ (J1-8)	VHF LOW B+ (J1-7)	VHF HIGH B+ (J1-6)	UHF B+ (J1-3)
VHF LOW (2-6)	+19V	+18V		
VHF HIGH (7-13)	+19V	+18V	+19V	
UHF (14-83)	+19V			+19V

Fig. 1-9. Band switching chart.

LED Display

The LED display decoder driver IC is used to drive a two-digit, seven-segment LED channel indicator. Eight lines of *binary coded decimal* (BCD) information are supplied to the IC located as part of the MSC control module to allow continuous, non-multiplex operation. The code which relates to the input and output states is the standard BCD-to-seven segment code which activates the appropriate segment or segments. Figure 1-10 gives the logic condition or output code of the LED display driver IC. If a problem occurs resulting in loss of, or incorrect LED display, Fig. 1-10 will aid in determining if the problem exists in the MSC module or LED assembly.

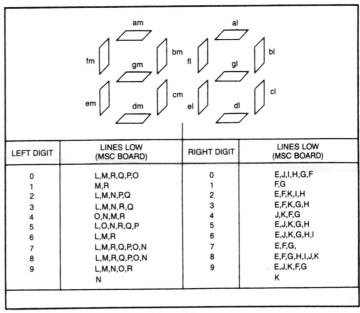

LEFT DIGIT	LINES LOW (MSC BOARD)	RIGHT DIGIT	LINES LOW (MSC BOARD)
0	L,M,R,Q,P,O	0	E,J,I,H,G,F
1	M,R	1	F,G
2	L,M,N,P,Q	2	E,F,K,I,H
3	L,M,N,R,Q	3	E,F,K,G,H
4	O,N,M,R	4	J,K,F,G
5	L,O,N,R,Q,P	5	E,J,K,G,H
6	L,M,R	6	E,J,K,G,H,I
7	L,M,R,Q,P,O,N	7	E,F,G,
8	L,M,R,Q,P,O,N	8	E,F,G,H,I,J,K
9	L,M,N,O,R	9	E,J,K,F,G
N			K

Fig. 1-10. LED display IC output logic condition table.

13

SERVICING THE SCAN FS CONTROL MODULE

In the following checks assume that the MST 001 tuner module and i-f link cable are operating properly.

Scan Control Module Symptoms

- Channel selecting problems in either VHF low, VHF high, or UHF bands.
- Channel selecting capability lost on all channels.
- Channel selecting OK-LED display is incorrect or segments missing.
- Improper or no volume-mute.
- No tuning voltages to MST 001 tuner.
- Channel selecting OK, but tuning system hunts.

Service Procedures

Check all interface connections and wiring to and from MSC control module. Make sure that channel entered is a valid entry. Each time a channel entry is made, note the display readout. This visual indication can prove to be a useful aid in tracking down a particular problem.

Channel Selecting Problems (VL, VH, or UHF Bands)

The above problems usually indicate a defect in either the MSC or MST modules. Check appropriate bandswitch voltages on the MSC module. If proper bandswitch voltage appears, sub in a new tuner module. Check the proper keyboard data lines that go to logic low when buttons are depressed on the keyboard. Logic low condition means voltage goes low, to about .2 volts or lower.

Channel Selecting Capability Lost on All Channels

First check that all dc operating voltages are present. Most dc voltages for the MSC 001 are derived from the negative 60 volt horizontal pulse. These pulses are scan derived from the sweep deflection system and can be checked with an oscilloscope. Next, check keyboard data lines; if ground is not available to the keyboard assembly, channel select capability is lost.

Channel Selecting OK-LED Display Not Correct

When the tuner responds to the correct commands and the display does not, check appropriate outputs from the MSC board. Note LED display output logic condition in Fig. 1-10 for data regarding logic condition of display driver IC located on MSC 001 module. Replace module if incorrect logic conditions are found on the MSC module. If the correct logic conditions are found on the MSC module, then the channel display assembly may be defective.

Improper or No Volume Mute

Check volume-mute line at J2 for a momentary "dip" toward a logic "low" condition which should exist when a channel change is initiated. If not, check plug connections and then try new MSC module. Voltage at J2 should be about +2 volts during normal station reception.

No Tuning Voltages to MST 001 Tuner

Replace tuner control module or sub in a battery and potentiometer, or power supply in place of the tuner voltage. Remember that proper bandswitch voltage must be present to perform this test. Refer to Fig. 1-8 for proper tuning voltages and prescaler outputs.

Channel Selecting OK-Tuning System Hunts

This type situation usually indicates an AFT problem. Tuning capability is not lost, but picture will drop in and out rapidly, or hunt. Check AFT voltage at J2 on the MSC module. Dc voltage should measure about +6 volts and should be stable with a strong TV station tuned in. Problem could be related to IF/AFT on main chassis or MSC control module. For other voltages and waveform information refer to the scan FS control module in Fig. 1-11.

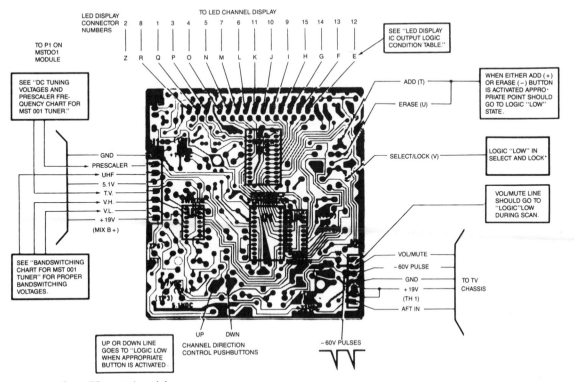

Fig. 1-11. Scan FS control module.

ELECTRONIC TUNING SYSTEM (VARACTOR) GE ET-20 SYSTEM

Three functional blocks make up the GE ET-20 system as shown in Fig. 1-12. The varactor tuner is designed to tune three separate frequency bands which are activated, one at a time, by the application of +22 volts to the proper tuner terminals. These bands are as follows:

Channels 2-6 54-88 MHz (low VHF)
Channels 7-13 174-216 (HIGH VHF)
Channels 14-83 470-890 MHz (UHF)

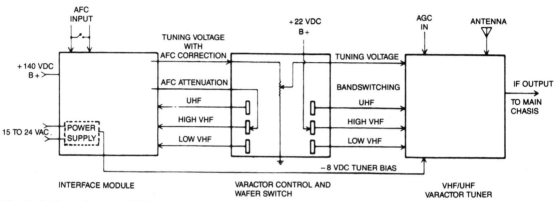

Fig. 1-12. Block diagram of ET-20 electronic tuning system.

The band switching voltage is fed by means of a wafer switch, controlled by the channel selector knob. The tuning of a specific channel within the frequency band selected is done with potentiometers in the varactor control assembly.

The varactor control assembly contains not only the bandswitching wafer switch, but also has twenty slide-type pots arranged on a cylinder. The wipers of each pot are mechanically engaged, one at a time, by an external tuning knob. The wipers are electrically connected to an output terminal on the assembly.

There are a different number of pots in parallel for each band. Five on the low VHF band, seven on the high VHF band, and four on each of the two UHF sections.

Each of the pots, once mechanically and electrically engaged, is capable of tuning in any channel within the selected frequency band.

The VHF channels are normally pre-tuned sequentially from the low to the high channels in order to match the numerals on the channel readout drum, and also to line up the red indicator with the white marks on the tuning dial.

The wafer-type bandswitch is mechanically connected to the potentiometer drum and, in addition to activating the proper tuner band with +22 volts dc, it also helps provide the system with the voltage needed to maintain the correct AFC range for each band. Each band requires a different amount of AFC voltage for 1 MHz of pull-in range.

The interface module shown in Fig. 1-13, contains the circuitry necessary to process the AFC and switching voltages required by the varactor tuner.

16

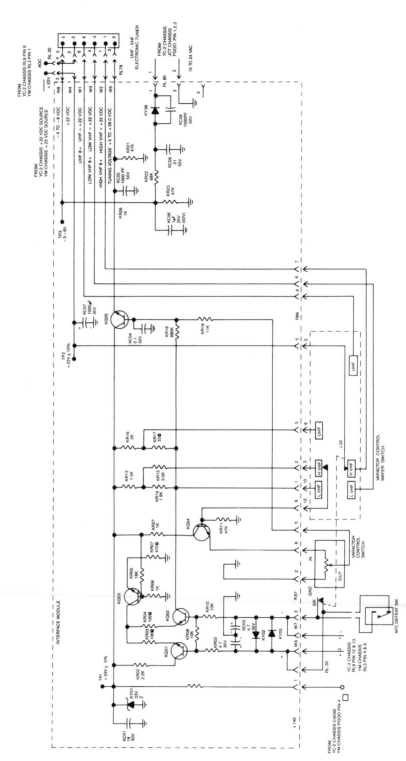

Fig. 1-13. Interface module diagram.

17

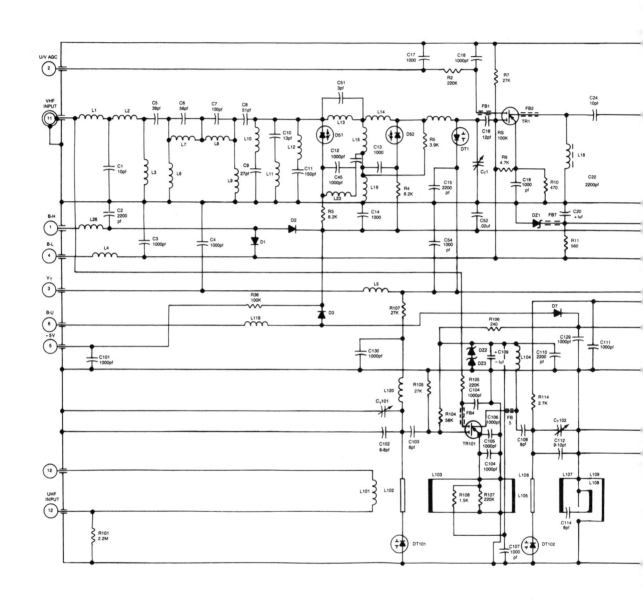

Fig. 1-14. GE Varactor electronic tuner circuit.

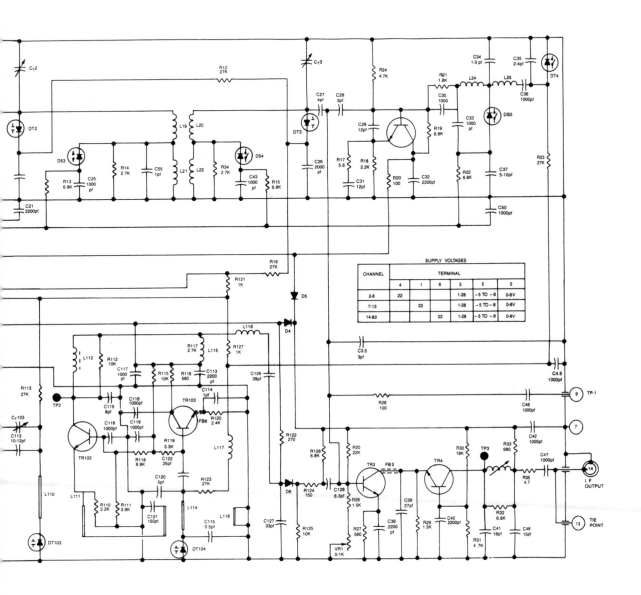

CHANNEL	TERMINAL					
	4	1	6	3	5	2
2-6	22			1-28	-5 TO -8	0-8V
7-13		22		1-28	-5 TO -8	0-8V
14-83			22	1-28	-5 TO -8	0-8V

SUPPLY VOLTAGES

19

The AFC voltage from the main TV chassis i-f module is applied to the network of KY02 and KY03. These diodes limit the differential voltage variations to 0.6 volts.

The differential amplifier stage KQ01, KQ02, and KQ03 modifies the AFC from a double ended to a single ended voltage source present at the collector of KQ02. Because of the requirement for AFC correction voltage decreases, from the low to high channels, three levels of attenuation are needed. These are provided by voltage-dividing resistors and are switched by the varactor control bandswitch to the base of KQ04. The divider network KR16 and KR17 is used for AFC level correction on the UHF band. KR13 and KR15 are used to attenuate the voltage for high VHF channels, while the collector voltage of KQ02 is used directly on the VHF low band channels.

The varactor control (tuning pots) tunes in the selected channel frequency by setting the voltage level required at terminal 3 of the varactor tuner via the buffer stage KQ05. The tuning voltage is altered by KQ04 in direct proportion to the AFC correction voltage present at its base.

A bias voltage which can vary from -5 to -8 volts dc is applied to terminal 5 of the varactor tuner. This voltage is developed by KY06 and its associated circuitry. The bias prevents the tuner from producing unwanted oscillations under certain conditions. The complete circuit for the VHF-UHF Varactor tuner is shown in Fig. 1-14.

ZENITH UNIFIED VHF/UHF TUNING SYSTEM

Zenith has developed another concept in electronic tuning systems. In the past, most tuning control systems have been separated from the tuners being controlled. The new 175-2253 system combines the functions of a VHF tuner, UHF tuner, and microprocessor tuning control system in a single module. It is mounted at the front of the TV set next to the keyboard and LED channel readout. The LED readout is mounted on the same printed circuit board assembly as the keyboard. A common bracket is used to mount these assemblies. Thus, the connecting wires between the assemblies are very short. The total number of interconnecting wires has been reduced. This greatly reduces the stray pickup of electric and energy generated by picture tube arcing.

This tuning system has several unique features. It has special fine tuning (the first time this feature has been included in a Zenith PLL Tuning system). The fine tuning circuit is of the non-memory type. A new 0.4 inch red LED Channel Readout has been used in place of the present 0.6 inch red or green display. There is also a new channel selection feature.

The VHF antenna input is 300 ohms, balanced. Capristors (2) are used on the antenna input terminal block at the rear of the TV set to provide the required UL isolation. The UHF antenna input is also 300 ohms, balanced. In this case, isolation is provided by the transformer type input circuit in the tuner.

TUNING CONTROL SECTION

This Zenith tuning system uses the 221-285 microprocessor. It is housed in a 28 pin package that allows the use of a more compact PC board as compared to the older 40 pin package. The PLL, LED driver, D to A converter for volume control, raster blanking, and sound muting functions are built into this new integrated circuit.

Another feature is the "Timed Entry" (One Button Channel Selection). After one or two channel number keys are depressed on the Space Command remote control unit (or on the set keyboard), the requested channel will automatically be tuned after a two second delay. The "Enter" key does not need to be depressed. However, if the "Enter" key is depressed, channel change will occur immediately.

Another new feature is manual fine tuning. To activate the fine tuning mode of operation, the ENTER button is depressed on the keyboard. The Channel Up and Channel Down keys then function as Fine Tuning Up and Down keys respectively. The LED readout displays the letter "F" during fine tuning. A range of ± 3.25 MHz can be tuned from the correct frequency. After a time of 4.5 seconds following the completion of fine tuning, the buttons on the hand remote unit (and on the sets' keyboard) will revert to the normal channel changing function. There is normally no need for fine tuning in systems where only off-the-air TV stations are received. Fine tuning is used when an MATV system provides the TV signals or when a video game or similar device is connected to the TV set. This becomes necessary because many of these signals are not transmitted exactly on the correct frequency.

VHF Tuner Section

In the VHF circuits abrupt junction varactor tuning diodes are used because they possess accurate and repeatable tuning curves. Every diode is matched to a specification, allowing better manufacture and service of the tuner, all of the diodes are very much alike. This feature allows the tuner to be used with a variety of tuning control systems.

The electrical performance is maintained using a smaller circuit area on the PC board. The 300-ohm antenna input now uses a Molex type plug for reliability and ease of service. Molex plugs are used for the i-f output and the power inputs.

UHF Tuner Section

Several design improvements characterize the UHF circuits in these tuning circuits. The design maintains high standards of quality and reliability. The electrical performance is as good as the 175-2031 and the noise figure is expected to be improved.

The UHF antenna input is accessible from the top of the tuner. The antenna connector has been changed to a 300-ohm Molex version, which is in contrast to the Brundy type used in the 175-2030 tuner.

A new and improved RF amplifier, FET 121-1103, has been used resulting in lower noise figures and better stability.

A modified oscillator circuit has been located directly on the printed circuit board. It was formerly mounted on a metal sub-chassis. A new oscillator transistor in a Micro-T package is now used. In order to further miniaturize the circuit, the resistor substrate network was replaced by discrete components.

The prescaler IC, 221-241, has been changed so that the oscillator frequency is divided by 64 instead of 256. It has been relocated to aid in the miniaturization of the tuner section of the system.

Because there is no interconnecting cabling between the tuner and tuning system, a potential source of problems has been eliminated.

MEMBRANE KEYBOARD

Some of the upper-end 25 inch "Z"-line television receivers will now feature a completely new keyboard system.

Instead of the traditional keyboard with the individual push buttons for channel selection and other functions there will now be a membrane keyboard in these receivers.

Membrane switches are reasonably new state-of-the-art devices that often replace conventional momentary-contact switches in low voltage, low current circuits. They are being used in high volume keyboard applications because of their relative simplicity. Compared to regular push button switches with springs, levers and pivots, most membrane types have only one moving part. Also, the membrane units typically need much less mounting depth than conventional switches.

The switch needs very little contact pressure and very slight switch travel. The membrane switches are usually less than .25 inch in thickness, and can be set close together and mounted on panels without the need for cutouts.

The contacts are hermetically sealed, they are protected against moisture, acids, corrosives, solvents, and dust for improved reliability.

Because membrane switch systems employ only one moving part they are not sensitive to vibration. Membrane switch pads are commonly specified for such high-use applications as telephone handsets, calculators, and terminal keyboards that demand as many as 10 million closures per switch. Membrane switches are commonly found on appliances, such as washing machines, dishwashers and microwave ovens.

Although the technology of membrane switches is not new, the use of a membrane keyboard is new to television and VCR recorders. A Zenith development program has led to the first application of a membrane switch keyboard to a TV tuning system. It will be integrated with other secondary TV controls in a compact convenient control center. The keyboard is located on the back of the secondary control door.

The keyboard itself is flat. There are no knobs or buttons. The various functions are outlined on the top overlay. This pattern is printed on the reverse side of this front layer of the keyboard assembly. The front of the overlay has an easily cleaned nonglare surface that is resistant to fingerprints and protects the membrane from sharp objects.

There are three layers in the actual keyboard. See Fig. 1-15. Two layers have an epoxy silver contact pattern for reliable electrical contacts. They are separated by a .005 inch thick layer with holes in each of the key locations. The layers are securely bonded to assure that the contact surfaces will be separated by the proper distance. The

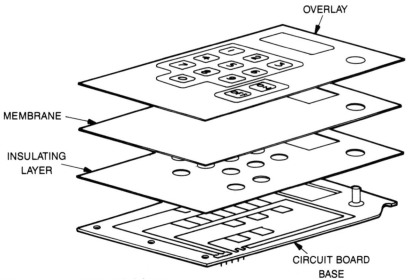

Fig. 1-15. Typical membrane switch layout.

contacts are connected by a ribbon cable pattern printed on the contact layer that leads out to a Molex connector. This connector interfaces the ribbon cable to the discrete wire terminal assembly used on the tuner control module. See Fig. 1-16.

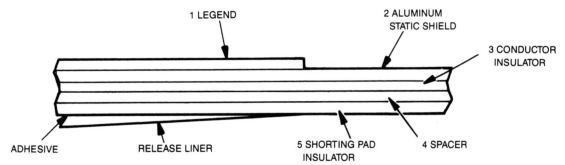

Fig. 1-16. Side view of membrane switch.

A unique feature of the new membrane keyboard assembly is the use of electrical shielding layers in front of and behind the actual keyboard contact layers. These layers consist of a screen grid pattern printed on an insulating material. They protect the keyboard and tuning systems from damage resulting from static electricity discharges. The shielding is also extended to cover the printed conductors leading to the Molex connectors.

The membrane keyboard is connected in a matrix. The microcomputer in the tuner assembly interrogates the matrix in order to determine if a contact has been made. And if a contact has been made, which function, then that function is executed and the microcomputer returns to interrogating the switch matrix, ready for the next contact closure.

23

2

TV Remote Control Systems

Several different manufacturers offer remote controls for TV receivers. This chapter covers some of the more popular models.

DIRECT ACCESS REMOTE CONTROL (ZENITH 2000 SYSTEM)

This remote control system will be found in the Zenith color TV set L models. This direct access remote hand unit transmitter is a 16 button keyboard unit. The functions include on, off, zoom, mute, volume higher and volume lower (continuous through 128 steps), plus channel entry buttons. The frequencies associated with the transmitter are in Fig. 2-1. A block diagram of the 2000 remote control system is found in Fig. 2-2.

Remote Hand Unit

The 124-14 hand unit transmitter is powered by a 9 volt battery. The 16 pin P-MOS digital IC is the heart of the transmitter. The IC senses any switch closures, scans the keyboard to determine which switch is closed, determines that the closure is valid, determines the X and Y coordinates of the closure, converts it to a 5 bit binary number which is used for gating internal counters to generate the desired ultrasonic frequencies. A simplified drawing of the hand transmitter is shown in Fig. 2-3. The output circuit is operated as a class C stage where pulses are applied to the tuned circuit of the ultrasonic speaker and output coil.

The best technique for testing a transmitter is to connect an oscilloscope to the microphone/preamplifier output. With the transmitter about one foot from the microphone, all buttons should provide a square wave signal which indicates the microphone/preamplifier is limiting the strong input signal.

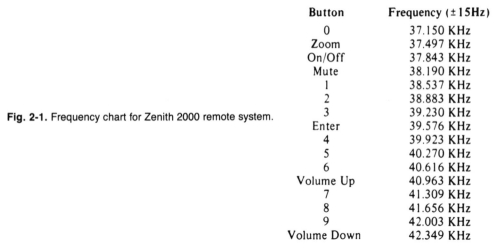

Button	Frequency (±15Hz)
0	37.150 KHz
Zoom	37.497 KHz
On/Off	37.843 KHz
Mute	38.190 KHz
1	38.537 KHz
2	38.883 KHz
3	39.230 KHz
Enter	39.576 KHz
4	39.923 KHz
5	40.270 KHz
6	40.616 KHz
Volume Up	40.963 KHz
7	41.309 KHz
8	41.656 KHz
9	42.003 KHz
Volume Down	42.349 KHz

Fig. 2-1. Frequency chart for Zenith 2000 remote system.

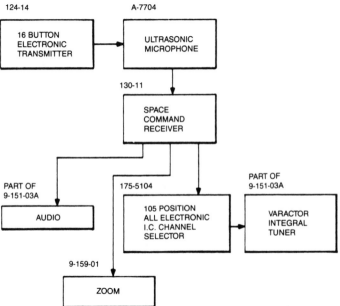

Fig. 2-2. Block diagram for Zenith 2000 remote system.

Microphone Amplifier Unit

This portion of the remote system consists of an ultrasonic microphone and a three stage solid-state amplifier. The wider bandwidth requirements of the direct access system have been accomplished by minute changes in the size of the microphone bridge. There is a range control located on the back of the unit that should *not* be adjusted to a *more* sensitive level than the factory setting but may be adjusted to a *less* sensitive level. Good shielding is required and the original shield cover must be retained in place. The B+ is provided directly from the +12 volt position regulator Q201 located in the remote receiver module.

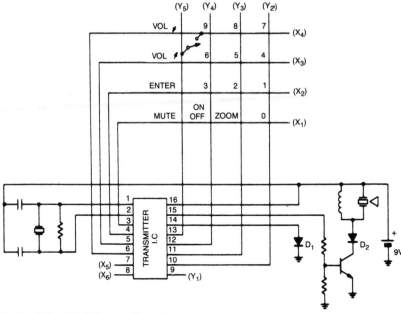

Fig. 2-3. Circuit for 124-14 hand unit. <small>(Courtesy of Zenith Corp.)</small>

Remote Control Receiver (130-11)

The direct access receiver is made up of two digital integrated circuits and 12 discrete transistors. Power is supplied by two series voltage regulators, Q201, a +12 volt supply, and Q202 a −6 volt supply. Both supplies will provide 40 mA of current. Follow along with this circuit operation by referring to the block diagram of the remote receiver in Fig. 2-4.

The digital P-MOS chip, IC201, processes the received ultrasonic signal and provides direct channel selection by interfacing with a microprocessor. It also controls auxiliary functions such as on/off, volume, mute and zoom.

Since the IC uses a 12 volt supply and the microprocessor a 5 volt supply, level shifters Q203, Q204, Q205, convert positive-going 5 volt scanning pulses from the microprocessor to negative-going 12 volt pulses on pins 11, 12, and 13 of the decoder IC.

A second IC, IC202, is a dual type C-MOS flip-flop. This is required as a memory storage band for the mute and zoom functions. This allows continuous operation of the mute and zoom functions with only momentary depression of the buttons. The outputs available from the decoder IC are high only as long as these buttons are pushed. Hence, the need for storage in a flip-flop.

Transistors Q209 and Q210 are used as current amplifiers to increase the small current capability of the C-MOS flip-flop to that required to pull in the zoom relay.

Another level shifter, Q207, changes the 5 volt 200 Hz clock signal, from the microprocessor, into a 12 volt signal. This is required to gate the mute function into the

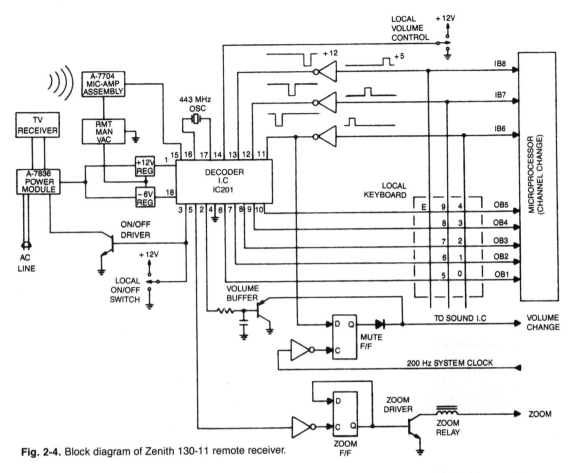

Fig. 2-4. Block diagram of Zenith 130-11 remote receiver.

mute flip-flop. Basically the microprocessor detects the mute function by scanning and then sends this function to be stored in the mute flip-flop.

Another inverter, Q208, is used to convert a negative-going transition occurring on pin 5 of the IC into a positive-going transition to clock in the zoom function.

A transistor, Q206, is used as a switch for activating a reed relay in the power module, which gates a Triac for turning the TV set on or off. This transistor can be changed from the off state to the on state either by remote control or manually by the local on/off switches.

The amplified ultrasonic signal from the microphone amplifier assembly is fed to the top end of R224, the trip point adjustment control. This control is factor set, using a standard 200 millivolt RMS signal to compensate for differences in individual chip sensitivity. This control should not be adjusted without proper instruments.

More System Operation Notes

Whenever an ultrasonic signal is fed into pin 15 of the IC201 (Fig. 2-5) a long chain of events begins to take place. After an ultrasonic signal has been received by the chip it

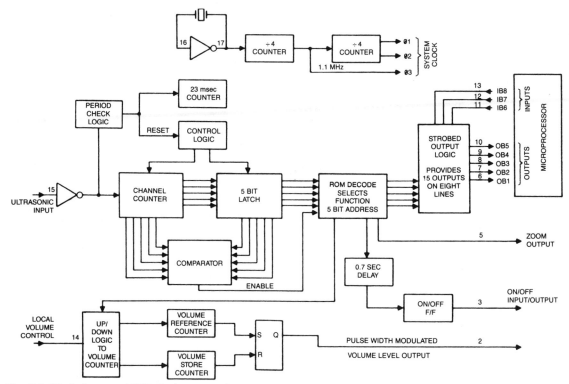

Fig. 2-5. Block diagram of IC201 in remote receiver.

begins a measurement cycle. During the first 23 milliseconds, the chip ignores (does not count) the signal and it is basically inactive. This is to allow any room transients to decay. However, the measurement cycle taken during the first 23 milliseconds determines if the received signal has a period which is longer than the minimum requirement of 18 microseconds yet shorter than 36 microseconds the maximum value of a valid signal. If the ultrasonic signal duration is less than 23 milliseconds or the period does not fall within the above limits the chip will not respond. If the ultrasonic signal has a valid period and is longer than 23 milliseconds duration a channel counter counts the number of ultrasonic pulses occurring during the next (2nd) 23 ms period. This value is then transferred to a 5 bit latch and the channel counter counts pulses for another (3rd) 23 ms period. At the end of the second counting period (3rd 23 ms period) the contents of the channel counter are compared to the previous value stored in the 5 bit latch. If a comparison does not exist, then another measurement is made, followed by yet another measurement, if there is again no comparison. Thus, there are three opportunities to obtain a comparison between two consecutive measurements. This is recognized as a valid ultrasonic input and the ROM decoder is enabled and the appropriate output appears.

A check on the period of the ultrasonic signal is made continuously. If, at any time, an ultrasonic pulse is received which has a period less than 18 microseconds or greater than 36 microseconds then the check period logic produces a signal which resets all control and measurement logic and inhibits the output stages.

The zoom output, pin 5 of IC201, is simply a buffered decode from the ROM. The on/off output, pin 3, is toggled by the appropriate ROM output after a delay of 0.7 seconds. The on/off pin is bidirectional since it also acts as an input pin. The external switches force the on/off pin to a particular state, the change is sensed by the chip and that level is held when the input is removed.

The pin 2 volume out is a 8.7 kHz square wave output with a variable duty cycle. At power-up the duty cycle is essentially 50 percent. This output is filtered by an integrator and applied to the base of the volume buffer to control the TV volume level. The duty cycle is controlled by two free counters, a reference counter and a store counter. A volume up or volume down command increments the reference or store counter respectively so that there is a resultant change in the duty cycle of the output signal. In order to prevent blaring when the set is first turned on, a separate adjustment control, off emitter Q212 connector 11, (factory adjustment) is provided to compensate for any sound i-f chip variations. The manual volume input pin 14 of the IC201 is also tri-state (like the on/off) and the volume is changed by forcing the pin to +12 volts or ground.

In order to conserve pins on the IC, 15 possible outputs are read via 8 pins. This is accomplished by arranging the pins as three input and 5 output scan lines. Refer to Fig. 2-6. If, for example, the desired function is f5 (digit 5) then by scanning, it is possible to read f4, f5 and f6 on one line i.e., OB4. The microprocessor does this by sending pulses separated in time such as t1, t2 and t3. During t1 only the END gates activated by t1 could be read out, namely f3 and f6. During t2 functions f2 and f5 may be read out. In this case f5 is high and this is shown as a high only during the time t2 is high. Since the microprocessor keeps track of which pulse is being sent, it knows which function the remote has decoded. By this means the digits 0 through 9, enter and mute are detected. As mentioned, mute is returned to be stored at the mute flip-flop during the negative transition of the 200 Hz system clock.

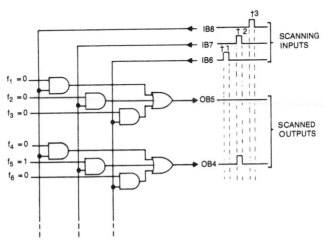

Fig. 2-6. Strobed output logic.

Service Tips

If set will not turn on check out the power supply module for the remote system. Check fuse, reed relay and Triac. Next, check for correct B+ voltages to the remote module. If set operates manually then see if the remote hand unit is transmitting the correct signals. If these checks are OK and the remote system seems dead check for the proper 200 Hz clock signal. Then suspect the microprocessor chip and the decoder IC, IC201.

MSC-003 FS SCAN REMOTE CONTROL (RCA)

The MSC-003 FS scan remote control module is used in the RCA CTC-93 color TV receivers. This unit works with the keyboard, remote receiver and preamp unit that makes up the total remote control system.

Control Module Symptoms

No or improper channel Up/Down action.
No or improper ADD or ERASE of scan memory.
No or improper channel change or skip.

Preliminary Service Information

Check all interface connections and wiring to and from the MSC control module. Each time the channel is changed note the display readout. This visual indication can prove to be useful in tracking down various remote control problems.

No or Improper Channel Up/Down Action

Scan up and down channel information requires that frequency synthesizer chip (U1) receive proper logic conditions to pins 16 and 17 (from scan channel switch). First, confirm good ground connection at terminals on the MSC-002 module and to up/down switch assembly. If ground connection is open, Scan capability is lost. If either up or down action is lost, check appropriate terminal on MSC-002 module for logic low condition when the correct button is pressed.

No or Improper Add or Erase of Scan Memory

Confirm that proper logic conditions at terminals T and U on the MSC-002 control module are being made. Make sure select-lock switch is in select position and that there is a good ground connection at all terminals on the MSC board. The appropriate add or erase line must go to logic low condition to indicate an add or erase function to the Scan Memory IC (U4). When add or erase functions are not activated, add and erase lines should be idle at logic high (+5 volts dc).

No or Improper Channel Change or Skip

These problems are usually associated with the MSC-002 control module. If quick voltage checks do not pin down the problem the best bet is to replace the MSC-002 control module.

The FS Scan Remote System (see circuit with voltage and waveform notes in Fig. 2-7) utilizes a frequency synthesizer chip (U1), Op/Amp chip (U2), Memory chip (U4), and an on-screen channel display/clock chip (U3). These IC's and associated components make up the MSC-003 tuner control module for the Scan Remote System. With this system, we will now look at the display/clock interface and other remote servicing procedures. Again, much of the circuitry in the system is similar to that found in the keyboard and scan systems. To locate problems, use the isolation techniques to help you pinpoint the portion of the system that is at fault.

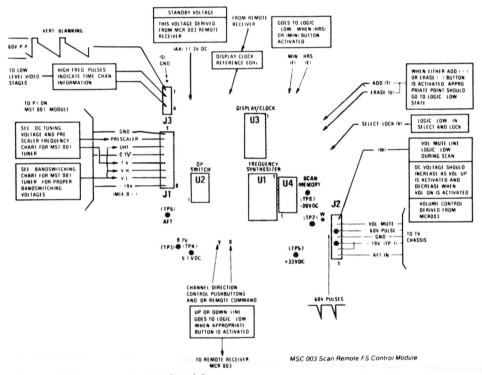

Fig. 2-7. RCA scan remote control module.

The Scan Remote System includes the MCR-003 remote receiver which utilizes the remote decoder IC that is not shown. This IC processes channel up/down, set on/off, and volume up/down information from the remote control unit via the preamp unit, and sends appropriate voltages to the MSC-003 control module. The remote receiver also processes function commands from the manual pushbutton assembly located on the TV receiver. Also, located on the remote receiver is a +11 volt dc regulator (Q1101), which provides dc voltages to the remote receiver and preamp board (as long as the TV receiver has ac power connected).

A separate power supply transformer (T1) located on the remote receiver assembly provides ac power (12 volts) to regulator Q1101 for standby and operation voltage to the remote receiver and preamp. Relay K1 provides on/off ac power to the TV receiver.

Remote Module Symptoms

Unable to set time (hours or minutes).
Erratic display, loss of time display.
No channel up/down—Remote.
No channel up/down—Manual buttons.
No channel up/down—Remote or Manual.
No or improper volume up/down—Remote.
No or improper up/down—Manual buttons.
No or improper volume up/down—Remote or Manual.
No remote control action.
No on/off action from either remote or manual buttons.

Service Checks

Check all interface connections and wiring to and from the MSC-003 control module, remote receiver MCR-003, Preamp MCY-003, and other assemblies associated with this system.

Unable To Set Time

Display problems can be defined as a loss of, erratic, or distortion on screen digit display with otherwise normal operation. Such problems are usually confined to the clock and display IC located on the MSC-003 control module or to connector problems. If these problems are encountered, replace the MSC-003 module. A defective display assembly may also cause instrument video problems on the sets screen. If video problems are suspected as being caused by the display system, remove connector P3-MSC and see if the problem has been cleared.

Failure of the system to maintain the correct time-of-day is a comparatively improbable situation without any other symptoms being evident.

Check If A Problem Is Encountered:

Check for possible intermittent power interruptions. If power to the TV set has been interrupted, the time-of-day displayed will be lost when power is restored—requiring the clock to be reset.

Check the time-set switches and cabling for any possible intermittent connections.

If the clock and display is not operating properly, then check for proper dc voltages around the pins of the U3 IC located on the MSC-003 module. If no reason for an incorrect dc voltage can be found then the U3 IC chip may be defective and will have to be replaced. If the chip is hot to the touch it may well be defective.

Remote Scan Turn On/Off Problem Notes

In the scan remote system the MCR-003 is used in the CTC-92 and 93 chassis and the MCR-004 is used with the CTC-97 chassis. Repair of the remote receiver is accomplished by the substitution of the remote amplifier module.

Because the CTC-92, 93 and 97 chassis are of the Extended-Life-design, a no-turn-on symptom can, of course, be a fault in the TV chassis rather than the remote receiver module. Thus, it is good to use a simple procedure to assist in locating the exact fault area.

The triac is not used in the FS Scan Remote Systems. Instead, a relay is used as the off/on switch.

The relay driver or switch in each case is a transistor, again controlled by preceding logic circuitry. The relay driver in the MCR-003 is a PNP device with the emitter connected to the 16 volt source through the relay winding. Thus, if the emitter is connected to ground with a clip lead, the TV set should turn on. The relay driver in the MCR-004 is essentially an NPN device so that the collector lead may be shorted to ground to turn on the set.

You may think the above procedure is not required because an audible click is heard from the relay when the on button is depressed. However, this is not always so. Don't forget that an audible click does not rule out the possibility of oxidized or otherwise defective relay contacts. Of course, this type of problem can be isolated simply by shorting the relay contact terminals 2 and 5 together. If the set does not turn on with this test, the problem is within the TV chassis.

REMOTE RECEIVER POWER SUPPLY (RCA CTC-48-68)

We will now look at some RCA remote control power supply circuits and service tips used in the CTC-48 through CTC-68 chassis series. These sets have the stepper system circuits.

Power Supply Circuit Analysis

As you will note in Fig. 2-8 there is nothing complex or unusual about this power supply circuit. The primary of the remote power transformer, T105, is protected from an overload by a .2 A fuse, F103. A metal oxide varistor, RV101 is across the ac line to protect the triac and other devices from any power line surges and spikes. You may want to add a MOV to any sets you find that do not have one in place, especially if you have had line surge damage to the triac and/or remote receiver devices. Addition of the MOV will help ensure against other component failures. Make sure you add the MOV after, and not ahead of, the fuse.

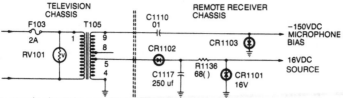

Fig. 2-8. Power supply circuit for remote receiver.

Approximately 12 volts ac from terminal 5 of the secondary of T105 is rectified by CR1105, filtered by C1117, and zener diode regulated to 16 volts dc by CR1101. The set cannot be turned on either by the manual off/on button or by the remote transmitter if this 16 volt dc supply is missing. If the supply is absent just check with an ohmmeter to confirm the presence of an open or shorted condition in either the rectifier or the Zener diode. If either of the devices is shorted, the .2 A fuse will be open. An ac voltage check at T105, pin 5, will reveal any possibility of a defective power transformer, although the line surge required to damage the device is usually very high.

One somewhat unusual symptom is that as long as the off/on button is held depressed, the set will remain on with distortion in the sound. As soon as the button is released, the set will turn off. An open C1117, 250 μF capacitor, is probably the cause of this symptom.

The top of the remote power transformer secondary, pin 9, delivers ac which is rectified and used to supply -150 volt dc bias to the high impedance remote receiver microphone or transducer. If this supply is not present, or very low, the sensitivity of the remote receiver will be greatly reduced.

Caution

When measuring this bias voltage, use a meter with a high input impedance like a FET or VTVM. A voltohmmeter will load the circuit and give a lower reading than the true voltage value. If the supply voltage is low or missing, possible causes are a leaky or shorted CR1103 or a defective remote preamp module.

REMOTE CONTROL LOGIC SYSTEMS (RCA CTC-74, 81 CHASSIS)

We will now discuss the stepper system used in several RCA color TV remote control systems. These electronic control systems contain gates which are used to allow remote manipulation of the on/off, channel change, volume level, color level and tint control range. Some combinations of these control functions are used in all RCA Remote Control Systems. This logic circuitry in the later models is modular in design so that most repairs can be completed with a simple module change.

One of the least complex and most widely used logic circuits found in the RCA Remote Systems is the volume stepper system. Because it is not a plug-in module and must be repaired by the technician we will look at its circuit operation and give some troubleshooting information. Now, follow along with the partial stepper circuit shown in Fig. 2-9.

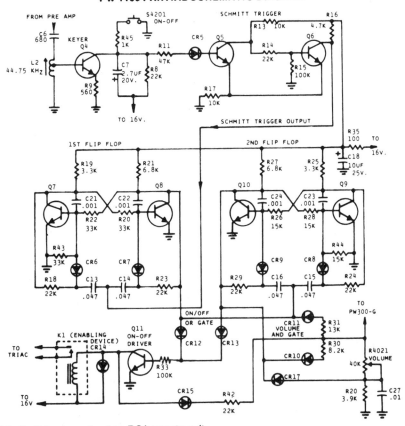

PW 1100 PARTIAL SCHEMATIC DIAGRAM

Fig. 2-9. Partial schematic of an RCA remote unit.

Volume Stepper

In the volume stepper, two flip-flops or bistable multivibrators controlled by a pulse from the Schmitt trigger ultimately determine the logic state of an OR gate which controls the on/off function of the main TV receiver. The OR gate consists of diodes CR12 and CR13, one or both of which must be on (forward biased) to hold the on/off driver, Q11, in saturation. When this transistor is held in saturation, the enabling device is energized to provide the triac gate pulse.

Another pair of diodes controlled by flip-flops, CR10 and CR11, comprise an AND gate which steps the volume through three levels. In series with each of the diodes is a resistor which is switched in parallel with the volume control when that particular diode is forward biased. The smaller the total resistance, the louder the volume. Therefore, when the set is first turned on, the volume is at maximum because CR10 if forward biased by the on condition of Q10 (collector is at logic low). Resistor R30, which is in series with CR10 is the smaller of the two. With the second push of the stepper switch, CR11 is forward biased by conduction of Q8 placing R31 across the volume control. Since the total volume control resistance is thus increased, the set is now at medium volume. With the third push of the stepper switch, Q8 and Q10 are off. Because neither resistor now parallels the volume control the TV sound level is now lowest (muted).

Again, referring to Fig. 2-9, we see the collector of the keyer transistor, Q4, may be taken low either by base bias provided by an ultrasonic signal from the remote transmitter or by manual activation of the stepper switch. The rather poorly shaped pulse produced, will reverse bias CR5 and turn off Q5, the first transistor in the Schmitt trigger. The second transistor in the Schmitt trigger, Q6, which is normally held off by the saturated state of Q5, turns on and a square wave is thus produced. The primary purpose of the Schmitt trigger is to square up the poorly shaped pulse from the keyer transistor. This square wave, of course, is used to provide gating for the first flip-flop.

The flip-flop is a bistable circuit which can exist in either of two stable states indefinitely and has a memory. When power is first applied to the flip-flop, the transistor that conducts first is the one with the larger collector load resistor. It will remain in that state with one transistor on and the other off until it receives a gate pulse. It will then switch states and remain so until the next gate pulse. Gating for the first flip-flop is provided by the Schmitt trigger. The gate pulse for the second flip-flop is provided *only* when the second transistor in the first flip-flop is *on*. Therefore, while the first flip-flop changes state with every push of the stepper switch, the second flip-flop changes state only every *second* push of the switch.

As shown in Fig. 2-9, CR12 conducts when Q8 is off (collector high) and CR13 conducts when Q10 is off. Therefore, the TV set remains on as long as either or both transistors are off.

From the preceding information, it appears that the entire system is a complex switching network. Indeed, any logic gate may be schematically represented as a switch. It must be remembered, however, that either logic high or logic low can be chosen as the *on* state.

Since the collector of any transistor in the system is at logic high when the transistor is *off,* the logic low condition may be simulated simply by shorting a certain collector to ground. Referring to the schematic in Fig. 2-9, you can see that when the TV set is off, Q4, Q6, Q7, and Q9 are all off. All things being normal, we need only to momentarily short the collector of one of those transistors to ground to turn on the TV set. Because of the memory characteristic of the flip-flop, the set will remain on until it receives another command. Using this clip lead method, it is possible to step the system through the four separate states shown on the chart. Also, this method will help you to determine whether a particular circuit—the keyer, the Schmitt trigger, or one of the flip-flops—is defective.

Turn On/Off Problem Notes

The entire remote control receiver circuitry in the Frequency Synthesis Scan System is contained on one of two modules. The MCR-003 used with the CTC92 and 93 chassis and the MCR-004 used with the CTC97 chassis. Repair of the remote receiver is accomplished, of course, by the substitution of the remote amplifier module.

The CTC92, 93 and 97 chassis are designed in such a way that a *no-turn-on* symptom can, of course, be a fault in the TV chassis rather than with the remote receiver. Thus, we will now look at a simple procedure to assist you in determining where the trouble is to be found.

The triac is not used in the Frequency Synthesis Scan Remote Systems. Instead, a relay is used as the off/on switch.

The relay driver or switch in each case is a transistor, again controlled by preceding logic circuitry. The relay driver in the MCR-003 is a PNP device with the emitter connected to the 16 volt source through the relay winding. Therefore, if the emitter is connected to ground with a clip lead, the TV set should turn on. The relay driver in the MCR-004 is essentially a PNP device so that the collector lead may be shorted to ground to activate the set.

If you hear a click of the relay this does not always mean that the relay is good, as the relay contacts could be oxidized and not make contact. However, this problem can be checked by shorting the relay contact terminals 2 and 5 together. If the set does not turn on with this test, the problem is then isolated to the main TV chassis.

MAGNAVOX "STAR" TV TUNING SYSTEM

This electronic tuning system has four main units: These are the keyboard assembly, the remote receiver assembly, the transmitter unit, and the tuning assembly. The main parts of the electronics are located on four modules: the power supply module, the logic module, the tuner control module, and the video clock module. Some models have a time clock display module. Note the block diagram for the Star System in Fig. 2-10.

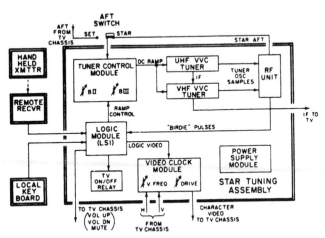

Fig. 2-10. Block diagram of Magnavox STAR Remote System.

One of fifteen different frequencies are fed from the hand unit transmitter to the remote receiver. The output of the receiver is then coupled to the logic module. A keyboard unit on the set can also be used to control the Star System. In either case, the logic circuits control all functions. The logic circuits are contained within LSI IC's. This custom-made device controls TV set off/on, volume level, channel changes and mute. A "logic" video signal, consisting of rectangular pulses, passes through the video clock module to the luminance circuit in the TV chassis. Horizontal and vertical deflection voltages from the TV set serve as timing pulses.

The tuner control module applies a dc ramp voltage to the voltage variable capacitors, or VVC, tuners. The LSI causes the dc ramp, or tuning voltage, to change when a new channel is selected and shifts the tuner oscillator to a new frequency. The tuner i-f output signal is then coupled to the i-f module on the TV chassis.

A sample of the tuner oscillator signal is coupled to the RF unit. The RF unit contains an internal crystal controlled oscillator which is used to produce "birdie" signals during the tuning cycle. The "birdie" signals are processed in the LSI and provide precision control over the dc ramp voltage and tuner oscillator frequency.

An AFT circuit in the RF unit provides precise correction of the dc tuning ramp in the tuner control module. The AFT correction circuit in the TV chassis may also be used if necessary. In most cases, the AFT switch should be operated in the *Star* position. The set position is used when a wider pull-in range is required to correct the larger carrier frequency errors found on some cable TV systems.

Video Clock and Time Display System

The circuits on the video clock module are involved with the development of the character video signal. These circuits consist of the character VCO ramp generator, the pulse shaping circuits, and the vertical multivibrator. The remaining circuits are located on the logic module and include the horizontal multivibrator and the character generator. Note block diagram in Fig. 2-11.

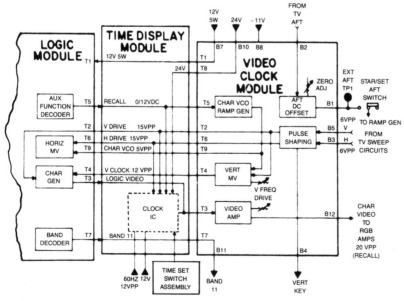

Fig. 2-11. Video clock and time display modules.

When the recall button is depressed or when a channel is selected, a dc voltage is applied to the character VCO ramp generator. The output goes high (5 V) and stays high as long as the button is held down or until the tuning cycle is completed. This voltage determines the frequency of the two voltage-controlled multivibrators and the frequencies determine the size of the characters displayed on the screen. When the recall button is released or the tuning cycle completed, the characters remain at maximum size for a moment or two and then decrease in height and width until they suddenly disappear. This reduction in character size is caused by the character VCO signal ramping down to zero voltage gradually, rather than cutting off abruptly. Timing pulses from the TV deflection circuits are shaped and applied to the multivibrators to key them off at the proper time.

When the vertical multivibrator is first turned on, it oscillates about 1 kHz. As the character VCO signal decreases slowly from maximum to zero voltage, the oscillator frequency increases to about 4 kHz. Maximum vertical size is produced at 1 kHz. As the frequency is swept upward to 4 kHz, the vertical size of the character diminishes until the oscillator is cut off. The frequency of the vertical multivibrator can be varied with the vertical frequency control on the video clock module. This control allows the vertical size of the channel number display to be adjusted.

Logic Module

Each key on the local keyboard produces a BCD number on the four output lines. These voltages are coupled through gates to the data decoder and the auxiliary function decoder circuits in the LSI on the logic module which is shown in Fig. 2-12. When the transmitter is used to operate the system, the signal is amplified by the remote receiver and fed to the remote decoder. The remote decoder portion of the LSI contains a "frequency window" circuit which blocks out extraneous signals outside the 34 kHz to 44 kHz frequency range. Special circuits "decode" the transmitter signal by converting specific frequencies to a corresponding BCD code. The output voltages parallel the BCD lines from the keyboard gates so that either the transmitter or local keyboard may be used to control the system.

The data decoder circuit accepts all BCD information relating to the selection of TV channels, and rejects all BCD data relating to the auxiliary functions. In a similar manner, the auxiliary function decoder accepts BCD information relating to the on/off, volume, mute, and channel recall functions, but it rejects all TV channel selection data.

Servicing the Star Remote System

The signal-flow block diagram in Fig. 2-13 will serve as a good troubleshooting aid in problem diagnosis. This diagram locates the major circuits and functions within the system. Signal flow between circuit functions and modules is depicted along with significant ac and dc voltages. The diagram can be used to determine which module might cause a particular trouble symptom and help pinpoint the problem.

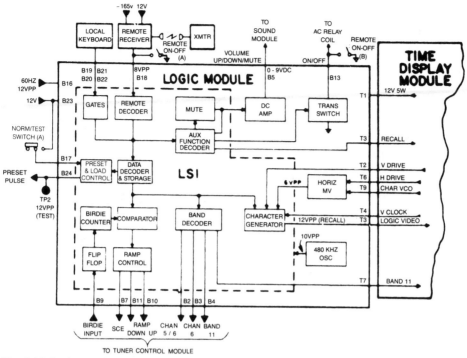

Fig. 2-12. Logic module found in STAR system.

Many of the problems that occur in these remote systems can be repaired on a service call by module replacement. As an example, the remote hand unit can be subbed with a good unit. The remote receiver unit and local keyboard assembly can also be checked by substitution. When problems occur in the tuning assembly, the modules can be substituted one at a time until the defective one is located. The power supply can also be replaced on a home call. Voltage checks should be made before the power supply is changed. Use the troubleshooting guide in Table 2-1 for suggested causes of various trouble symptoms.

Table 2-1. STAR Troubleshooting Guide for In-Home Service.

1. Transmitter has no effect: Keyboard operates normally.

    a. Remote Defeat Switch in off position.
    b. Transmitter battery dead.
    c. Sub Transmitter assembly.
    d. Sub Remote Receiver assembly.
    e. Sub Logic module.

2. Local Keyboard inoperative or erratic: Transmitter operates normally.

    a. Sub Local Keyboard assembly.
    b. Sub Logic module.

Table 2-1 Continued

3. Neither Transmitter nor Local Keyboard operates normally.

    a. Sub Transmitter assembly.
    b. Sub Remote Receiver assembly.
    c. Check for external signal, such as a Security Alarm.
    d. Sub Local Keyboard assembly.
    e. Sub Logic module.
    f. Replace Power Supply module. Check dc voltages before changing.

4. No control of Volume or Mute; other functions normal.

    a. Sub Audio module on TV chassis.
    b. Sub Logic module.

5. No Channel Recall; other functions normal.

    a. Sub Video Clock module.
    b. Sub Time Display module.
    c. Sub Logic module.

6. Problems with time display; channel characters may not be displayed.

    a. Sub Time Display module.
    b. Check Time Set switch assembly connection.
    c. Check P111 connection on Time Display module.

7. TV set will not turn on with Transmitter or Local Keyboard.

    a. Master Power switch not on.
    b. Place Remote on/off switch in off position.
       (1) If set comes on, replace Logic module.
       (2) If set does not come on, check line voltage, line cord, Master Power Switch, circuit
           breaker, module and plug connections, on/off Relay, Power Supply module.

8. Random shutoff of TV set.

    a. Momentary loss of ac power.
    b. Check the ac input circuit.
    c. Modules or plugs are not seated firmly.
    d. Check for intermittent contacts on the Top Interconnect board and the Bottom
       Interconnect board. .Clean male pins with Magnavox Contact Cleaner/Lubricant spray,
       Part No. 171378-1.
    e. Sub Logic module.
    f. Change Power Supply module after checking dc voltages.

9. Stations are mistuned; other functions normal.

    a. Place AFT switch in the "STAR" position.
    b. Cable TV signal off frequency. Place AFT switch in "Set" position.
    c. Check Band adjustments.
    d. Sub Tuner Control module.
    e. Sub Logic module.

10. Snowy picture.

    a. Check setting of Sharpness control.
    b. Check antenna system.
    c. Check setting of RF Delay control on TV chassis.
    d. Check AGC voltages on TV chassis.
    e. Check RF AGC voltage on tuner.
    f. Sub Tuner Control module.

On the Bench Service

When in-home checks will not locate the fault, then more troubleshooting checks are needed on the shop bench. The main components that become suspect are the VVC tuners, the RF unit, and the interconnections within the tuning unit and between the tuning assembly and TV chassis. Some faults may be in the main TV chassis and not in the Star System. For example, a symptom of no picture, no sound or snowy picture may be caused by a defective i-f stage or AGC circuit on the TV chassis as well as a VVC tuner problem.

ZENITH SPACE COMMAND

The Zenith SC3400 Remote Space Command uses the 221-246 microprocessor, which has a built-in character generator.

The character generator system utilizes a sandcastle (composite sync) pulse input fed directly to an input pin on the microprocessor or on the character generator IC. The signal is internally separated into the required horizontal and vertical components. Horizontal information is derived from a counter that places the display box on the screen in the location determined by the setting of a setup control.

The character driver circuit has been considerably simplified. It allows the use of yellow characters with sharper edges. Snivets and other interference to the display has been greatly reduced.

Two universal field replacement modules (9-239 and 9-240) are capable of operation with both the sandcastle input and the conventional horizontal and vertical inputs. It can also display characters on the screen in either white or yellow. This is a money saving service feature since servicing is simplified and only one module type need be stocked.

An improvement in reliability results from the use of a single 5 volt regulator and the reduction of interface components, both active and passive.

A new character generator IC is used in Space Phone TV receivers. It is a revised version with the same performance as earlier model units. It accepts composite sync pulses instead of horizontal and vertical interval information. The on-screen display characters are yellow, presenting a different, better looking appearance.

Zenith Remote Transmitters

The slim line transmitter remote system uses the new 221-242 CMOS IC and a new "J" type alkaline battery. This new battery will last three or four years under average use. The new "J" battery provides a more uniform voltage and higher current. It allows one capacitor to be eliminated from the transmitter circuit, which allows a thinner design of the overall transmitter case. The new transmitters are more reliable and easier to manufacture.

The new IR diode package was selected because it converts input current to IR light output more efficiently. A new driver transistor has been chosen to present a more consistent drive current to the IR diode.

The combination of these improvements allows one of the new high efficiency diode packages to replace three of the diodes used in previous transmitter design with no decrease in the effective transmission range of the unit.

Another new part is a 640 kHz ceramic resonator, which replaces the R-C oscillator used with the old 221-157 series of transmitter ICs. The new resonator is more reliable and eliminates the need for factory adjustment of the oscillator circuit.

IR Transmitter 221-242 IC

Initially the X port (pins 1-5 and 16-18) is set to a logic high output condition and the Y port (pins 7-10) is set to the input state. The remote transmitter circuit is shown in (Fig. 2-14).

When a key is depressed on the transmitter keyboard, a connection is made between the X Port and the Y Port. The Y Port senses a key closure. The oscillator is turned on and runs for 37 milliseconds to allow for key debounce; after the 37 ms, if the X-Y closure still exists the Y Port senses the column in which the key is located. This information is stored in the X-Y decoder of the IC.

The X Port is switched to the input state. Now, the Y Port is placed in the output state at a logic high level so that it can be sensed at the X Port (which remains set to the input state). The row in which the depressed key is located is sensed at the X Port and stored in the X-Y decoder and the Key scan sequence is complete.

At this point, the X-Y Discriminator determines which key has been depressed and causes the appropriate command information to be transmitted to the TV set.

The new printed circuit board, which has been designed to fit into the new series of "Slim Line" Space Command transmitters, is common to all models. Only the keyboard overlay and the external keys change between various models. The use of a common PC board results in more efficient factory production.

Transmitters for SC2400 models are carried over from the "Y" line. They continue to use a 9 volt battery. The three IR diodes are replaced with the same single package diode used in the Slimline series. The IR receivers are the same as those used in the previous model line.

Zenith Remote Antenna Switch

The four position remote antenna switch is a selector of signal sources for the antenna input of "Z" line Zenith remote control TV sets that have on-screen displays. As many as four separate signals can be connected to the switch box. The switch is remotely controlled with the remote transmitter or a button on the control panel.

Specifications

The remote antenna switch has four input ports and two output ports. The output ports are labelled "TO TV SET" and "TO CABLE CONVERTER INPUT". These ports are shown in the basic Switch Box diagram in (Fig. 2-15). The switch is remotely

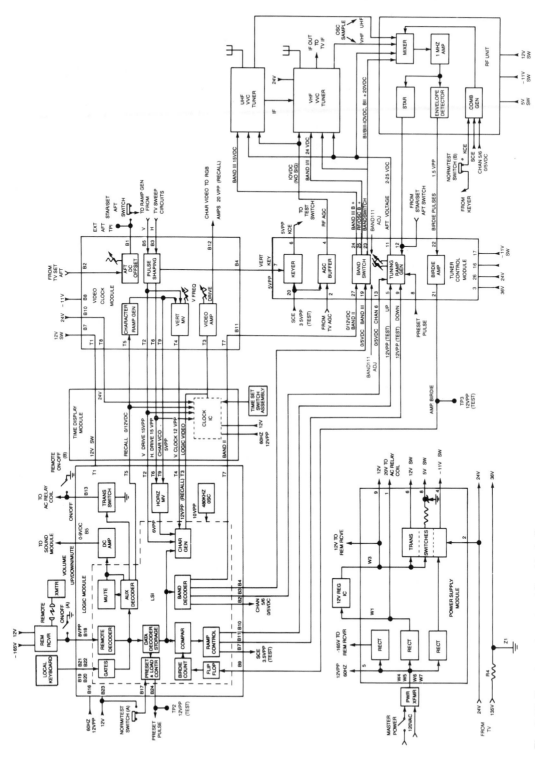

Fig. 2-13. Remote system signal flow block diagram.

44

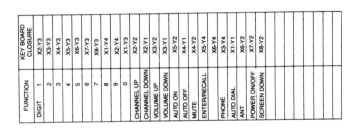

FUNCTION		KEY BOARD CLOSURE
DIGIT	1	X2-Y3
	2	X3-Y3
	3	X4-Y3
	4	X5-Y3
	5	X6-Y3
	6	X7-Y3
	7	X8-Y3
	8	X1-Y4
	9	X2-Y4
	0	X1-Y3
CHANNEL UP		X2-Y2
CHANNEL DOWN		X2-Y1
VOLUME UP		X3-Y2
VOLUME DOWN		X3-Y1
AUTO ON		X5-Y2
AUTO OFF		X4-Y1
MUTE		X4-Y2
ENTER/RECALL		X6-Y4
PHONE		X3-Y4
AUTO DIAL		X1-Y1
ANT		X6-Y2
POWER ON/OFF		X7-Y2
SCREEN DOWN		X8-Y2

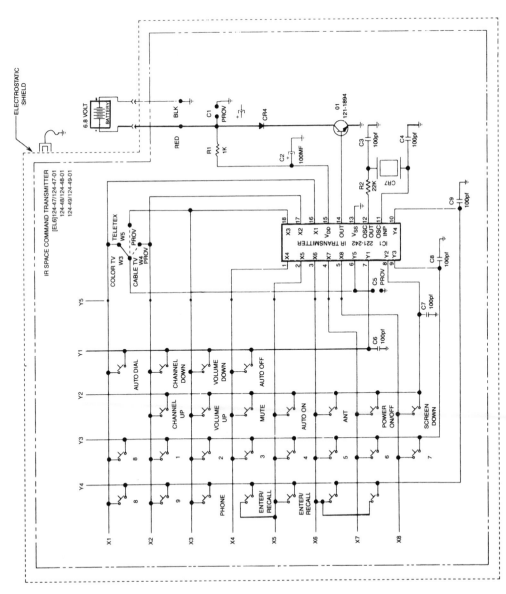

Fig. 2-14. Zenith remote transmitter schematic.

45

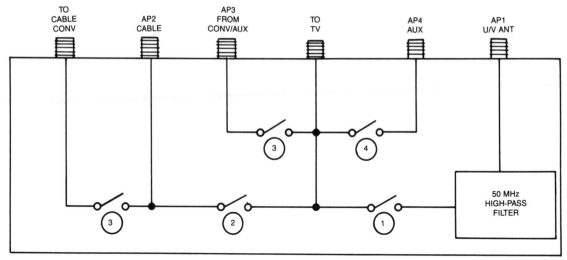

Fig. 2-15. Basic antenna switch box layout.

controlled. Any of the four inputs can be directly routed to the TV receiver using the antenna switch (ANT) button on the remote control transmitter. For example, the viewer could choose between an air signal from the antenna, a video game, and a cable system. The ANTENNA input has the capability to accept both UHF and VHF TV signals, CATV and a second CATV signal. At the CABLE input, the cable TV signal can go directly to the TV set or to a cable TV Converter/Decoder and then loop back to the TV set through the CABLE CONVERTER Input.

The Antenna Switch is designed to be mounted out of sight on the back of the TV set. The switch housing is electrically isolated from the TV set. The Remote Antenna Switch is shielded from Electromagnetic Interference (EMI) so that the video quality is not affected by RF interference from external sources. In addition, the unit contains a high-pass filter designed to reduce Citizens Band and Amateur Radio interference.

Switch Operation

When the Antenna Switch button on the remote transmitter is depressed, the identification letter corresponding to the selected Antenna position is displayed on the TV screen. Depressing the same button (within 4.5 seconds of the last button), will cause the next switch position to be selected. On Z-line TV sets with the space phone, "antenna" will be spelled out instead of "AP".

When the TV is turned off, it will remember the last Antenna Switch Position that was selected. The Remote Antenna Switch box has no power on/off switch because its power consumption is very small (3 watts). Power is obtained from a wall adapter, which remains plugged into the ac wall socket at all times.

All RF signal connections are made through 75 ohm "F" type coaxial connectors. To ensure 60 dB isolation between ports, 75 ohm terminations should be placed on all unused ports.

SWITCH IMPLEMENTATIONS

On the RF Switch printed circuit board, there are 25 high frequency switching diodes that are biased by 2.7 k ohm resistors and coupled by 22 1000 pF capacitors and one 2.2 pF capacitor. A 1000 pF feed through capacitor is used on the 26 V power supply line to the RF switch. The control lines from the "Z" line TV set are soldered directly to the printed circuit board. There is a 6.2 V zener supply for the CMOS decoder IC, which is used to turn on and off the five transistors that control the biasing voltages for the PIN diodes. There are two optical couplers for coupling the two dc command signals coming from the "Z" line TV set. Therefore, a 1500 V isolation exists between the "Z" line TV set ground and the Remote Antenna Switch ground.

Circuit Description

The circuit for the Remote Antenna Switch is done with PIN diodes connected in "T" type configurations. There are four control lines that control five basic single-pole single-throw diode switch configurations. The CONV-IN input switch and the CABLE TO CONV-OUT switch are operated by a common control line.

A PIN diode disconnects the two lower frequency ports when in the Cable or Antenna mode. This improves the frequency response of both higher frequency ports.

The control line voltage swings above and below the 6.2 V zener reference so that a single control line is used to forward and reverse bias the PIN diodes within a "T" type configuration.

All diode branches were designed for a 1.7 mA current drain when the dc power line voltage is 26 volts. The ac power line voltage will be 112 Vac when this condition occurs. During typical operation, the current drain will be in excess of 1.7 mA. If the ac power line is 135 volt ac, the dc power line is at 34.3 Vdc and current through a branch is 2.2 mA. At low power line voltage (90 Vac), the 26 V line voltage is 22.8 V and the current through a branch is 1.5 mA. Thus, there is no loss of performance at either low or high line voltage.

Therefore, a voltage regulator is not required. The total current drain at 120 Vac is 53.8 mA. The power consumed by the Antenna Switch Box is 1.74 watts.

ELECTRICAL DESIGN

In the electrical design, care was taken to remain below the rated maximums for all components by a factor of two or more. A high pass filter was placed at the antenna input. This filter helps suppress amateur radio and C.B. interference, and also decreases the possibility of lightning damage. Any surge of lightning energy will bypass the TV set and be routed instead to the ac power service. A resistor is used between the antenna input and ground to drain any static electric charge that might develop on the antenna. At the cable input, there is a shunt coil to ground and a 100 pF, 100 volt capacitor in series with the line. Together, they suppress the chance of lightning damage and discharges that may occur on the cable lines. The "1, 2, 3, 4" remote switch does not allow any significant amount of RF signal to travel back out onto the cable line.

3
Color TV Circuits

This chapter covers the various circuits found in a color TV receiver. A description of each circuit followed by the diagnosis of symptoms and troubleshooting checks will help you first identify problems, and then effect repairs.

HORIZONTAL OSCILLATOR/VERTICAL COUNTDOWN SYSTEM

This circuit is found in RCA's CTC 99/101 color TV chassis. The circuit uses an integrated circuit (a computer-related chip U400) to perform the operations of horizontal oscillator and vertical countdown. This combined countdown circuit feeds both the horizontal drive for sweep output and the vertical drive for the vertical amplifier. Note the block diagram of the countdown circuit in Fig. 3-1.

A horizontal rate output pulse is fed from pin 10 of U400 to the horizontal buffer Q406. The buffer stage then drives the driver stage and then the horizontal sweep stage. A vertical rate output pulse is available at U400, pin 8 (TP 9) and is applied to vertical sweep transistor Q500. The vertical pulse triggers the sawtooth transistor developing the ramp signal for the vertical amplifier.

Composite sync is fed to the horizontal oscillator/vertical countdown circuit via terminal S. From this point it is differentiated to provide horizontal sync pulses to pin 3 of U400. The sync pulses can be checked at TP-10. The composite sync signal is also coupled through two stages of integration to develop vertical sync at pin 5. The two stage integrator allows this vertical countdown system to be compatible with most forms of nonstandard sync signals.

The oscillator/countdown circuit is powered from a +9.3 volt regulated supply. The regulator is located within the IC and is biased through a resistor from the +23 volt

48

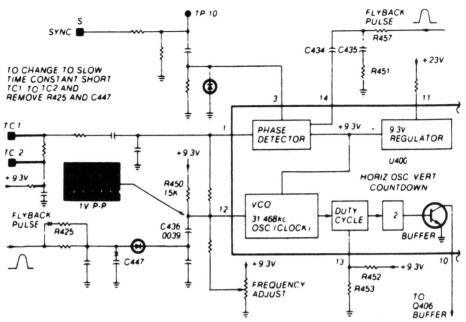

Fig. 3-1. Horizontal countdown block diagram.

supply. Pin 11 of U400 is the source for +9.3 volts to circuits external to the chip.

Controlling the horizontal oscillator/vertical countdown IC operation is an internal *voltage controlled oscillator* (VCO) which develops the clock frequency for the IC. The clock runs at twice the horizontal frequency for reasons we will see a little later on.

The VCO in the IC oscillates at 31.468 kHz, which is exactly twice the horizontal rate. The frequency of the VCO is determined by R450, C436, and the frequency adjust control. The VCO frequency is modified by the phase detector, which is located in the IC. The output of the phase detector at pin 1 is filtered, then coupled to the input of the voltage controlled oscillator (pin 12). To correct this picture bending, the voltage at pin 12 is modulated by feeding a detected sweep error signal to a (.0039) capacitor, C436. This is similar to "anti-bending" circuits used in other chassis.

The output of the phase detector modifies the VCO to achieve frequency lock to the station's horizontal sync signal. The phase detector compares the frequency of the horizontal retrace pulses to that of the incoming horizontal sync pulses coupled from the Sync/AGC circuit in the i-f. The horizontal sync pulses are differentiated and fed to pin 3 of U400. These pulses are then compared with the retrace pulses which have been integrated by R457 and C435. The resultant sawtooth is coupled to pin 14. The phase detector output dc voltage (pin 1) is fed to the VCO control input at pin 12.

The output of the VCO is coupled to a duty cycle adjustment circuit. The duty cycle is determined by a fixed dc reference voltage at pin 13 of U400. The output of the duty cycle block is applied to a divide-by-two counter, in which VCO output frequency is divided by two to provide the horizontal driver with 15.734 kHz horizontal pulses. The output of the counter is coupled through an internal open collector buffer stage to pin 10.

At this point, the output voltage square-wave varies between zero and about 4 volts. This square-wave is fed to the horizontal buffer stage which amplifies the signal which is then fed to the horizontal driver stage.

Vertical Countdown System

The vertical countdown concept is based on the exact relationship between horizontal and vertical sync frequencies. There are exactly 262.5 horizontal lines per field. Division (countdown) by other than whole numbers is not possible in this system; therefore, division of the horizontal rate by 262.5 to provide the proper vertical-to-horizontal sync ratio is not possible. To allow for countdown by a whole number, it is necessary to double the horizontal rate of 15.734 to 31.468 kHz. The result is a ratio of 525 to 1. The master clock VCO in U400 oscillates at the required 31.468 kHz rate. The clock output is then divided by two to provide the horizontal driving pulses and by 525 to obtain the correct vertical pulses.

The vertical countdown system used in the RCA IC is a two mode system. One mode is the countdown operation, the other is for sync operation. The two modes of operation allow the system to be compatible with nonstandard sync signals. The IC internally switches between the two modes of operation to match type of signal being received. Mode switching occurs after eight consecutive cycles of received vertical sync that is in phase, or out of phase, with the 525 count from the countdown system.

Standard NTSC Countdown Mode

The countdown IC uses counters, shift registers, coincidence gates, flip-flops, and other logic circuits to perform the necessary functions for countdown. The operation of the vertical countdown circuit operating in the countdown mode with an NTSC signal will be covered next. Refer to the vertical countdown block diagram in Fig. 3-2.

A 10-stage counter counts every clock pulse from the 31.5 kHz VCO. After counting 512 clock pulses, the greater than or equal to 512 count line goes high or to a logic 1. This places a logic 1 at one input of an AND gate. The other input of the AND gate remains at zero until vertical sync occurs. When vertical sync occurs, both inputs to the AND gate are ones, providing a 1 at the output which is supplied to one input of the coincidence gate. The other input of the coincidence gate is received from the 525 count line of the 10-stage counter. With the presence of an NTSC sync signal, the 525 count goes to a 1 at the same time vertical sync occurs. This provides a 1 at both inputs of the coincidence gate causing a 1 to occur on the yes line of the coincidence gate. This 1 is applied to the clear line of a 3-bit counter (clearing the counter) thus keeping the toggle flip-flop from changing states.

In addition to feeding an input to the coincidence gate, the 525 count line data is also supplied to one input of a two-input OR gate. The output of the OR gate is coupled to an R-S flip-flop causing the output of the flip-flop to change state (set) generating a vertical pulse via the output buffer stage. The pulse to the output buffer stage is also counted by the 3-bit counter. The logic 1 pulse at the output of the OR gate is also

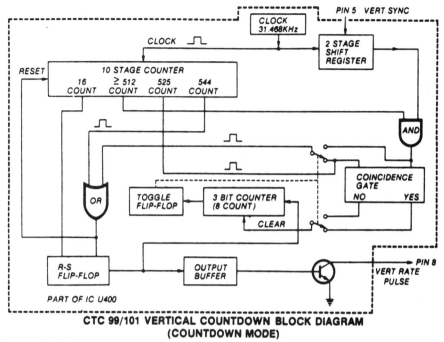

Fig. 3-2. Vertical countdown block diagram.

coupled back to the reset line of the 10-stage counter forcing it to reset to a 0 count.

The 10-stage counter again begins counting the clock pulses. After it has counted the 16th clock pulse, the 16 count line goes to a logic 1. This line is connected to the R-S flip-flop, resetting the flip-flop; thus, the length of the vertical output pulse is 16 clock counts. As long as an NTSC signal is being received, the operation of the countdown remains in this mode. Therefore, on every 525th count, the countdown circuit checks for coincidence of the 525 count and sync. As long as coincidence occurs, the coincidence gate clears the 3-bit counter, allowing the 525 count to trigger an output pulse through the OR gate and R-S flip-flop, thus, completing the cycle.

If no vertical sync is being received by the countdown IC, the second input to the AND gate remains at logic 0. This means that the one input to the coincidence gate stays at logic 0. Therefore, when the 525 count is reached and applied to the coincidence gate, the coincidence gate develops a 1 on the no line and a 0 on the yes line. This will clear the 3-bit counter. As before, the 525 count line energizes the R-S flip-flop through the OR gate. The flip-flop again drives the output buffer stage and also sets a 1 into the 3-bit counter stage. Also, the output of the OR gate resets the 10-stage counter to 0. Again, after 16 counts, the R-S flip-flop is reset, terminating the vertical output pulse. This cycle continues until the 3-bit counter counts eight occurances of no vertical sync occurring which is indicated by lack of a good pulse on the yes line of the coincidence gate. After receiving the eight consecutive pulses without being cleared, the 3-bit counter energizes the toggle flip-flop which shifts the mode of operation from the countdown mode to the sync mode.

In the sync mode, vertical scan is initiated by the occurrence of vertical sync. If no vertical sync is present, the countdown circuit will run free. The OR gate, which triggers the output generating R-S flip-flop, has two inputs: the output and the AND gate, and the 10-stage counter 544 count line. With no incoming vertical sync present, the AND gate will generate no output pulses. Therefore, the OR gate will be activated by the 544 count line only.

Horizontal/Vertical Countdown Circuit Servicing

For the following service information refer to the circuit and scope waveforms shown in Fig. 3-3.

The symptoms:

- No horizontal oscillator operations.
- Lack of horizontal or vertical sweep.
- Erratic horizontal operation.
- Erratic vertical operation.
- No vertical scan.

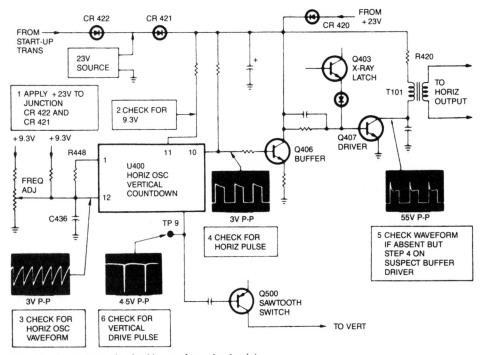

Fig. 3-3. Countdown circuit with waveform checkpoints.

The Diagnosis

Without the horizontal oscillator in operation, no other circuits in the TV chassis will operate, hence, there is no secondary B+ supply. To check the operation of the horizontal oscillator, it is necessary to apply an external source of +23 volts to the cathode of CR422 with the set turned off. Use a battery to sub for this power supply voltage.

Service Procedure

Use an isolation transformer, disconnect the line cord during all static checks, and use insulated tools or clip leads for any dynamic checks.

If during servicing it is necessary to replace the horizontal/vertical IC (U400), a resetting of the horizontal frequency adjustment may be required. To reset this adjustment, short the sync input (TP 10) to ground and adjust the frequency control for a stable horizontal picture.

- No horizontal oscillator operation.
- Lack of horizontal or vertical sweep.
- Erratic horizontal operation.

With +23 volts applied to the cathode of CR422, check for approximately 9.3 volts at pin 11 of U400. This voltage is internally regulated by U400.

Check for the horizontal oscillator drive output waveform at pin 10 of the U400 chip.

Check at the collector of the horizontal driver transistor Q407 for the correct drive waveform of 55 volts p-p. If the waveform is present at pin 10 of U400, but is not present at the collector, suspect a shorted Q407 transistor. If these circuits all check out good, then the problem may be in the horizontal start-up circuits.

Erratic Vertical or No Vertical Scan

Check for the presence of vertical drive pulses at test point TP-9. If the vertical pulse is present, the fault will probably be in the setup switch or vertical amplifier.

VERTICAL SWEEP AMPLIFIER (SOLID-STATE) RCA 99/101 CHASSIS

The main B+ power for this sweep system is a +23 volts instead of a negative supply that has been used in earlier model SS RCA sets. The retrace-switch transistor is supplied with +55 volts providing the extra required B+. The base bias supply for the top and bottom output transistors and the retrace switch is developed through a resistor from the +210 volt supply.

Brief Circuit Analysis

This vertical sweep circuit employs a vertical height tracking circuit. Height tracking is accomplished through R506, a 4.7 M resistor, which is connected to the beam current detector via the X-ray protection/shutdown circuit. Q501's base bias is modulated by the beam current, causing the vertical amplifier output to "track" with changes in average brightness. This maintains the proper aspect ratio between horizontal and vertical scan, preventing vertical "shrink" under high beam current conditions.

Vertical Amplifier Trouble Symptoms

- No vertical deflection
- Reduced scan
- Retrace lines at top of picture

Some Preliminary Checks

Pull setup switch in and out a few times to make sure that the switch contacts are not dirty. Check for presence of vertical drive pulse at TP 9. Check for correct B+ supplies: +23 volts, +55 volts, and +210 volts. The best way to service this vertical amplifier circuit is to place the service switch in the setup position. This prevents the oscillator stage from having any effect on the vertical amplifier circuit.

Troubleshooting Procedures

Use an isolation transformer and disconnect the line cord during all static checks. Refer to the simplified diagram in Fig. 3-4 with the boxed service checkpoints.

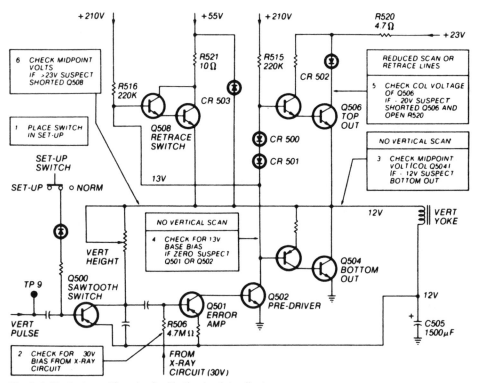

Fig. 3-4. Vertical amplifier circuit with checkpoint callouts.

No Vertical Deflection

Check the Q501 error amplifier for a 30 volt bias supply at the base. This voltage is developed in the X-radiation protection circuit and can be measured at the end of R506 away from the base of Q501. Without this voltage, the vertical amplifier will not operate. If this voltage is not present, check out the X-radiation protection and shutdown circuitry.

Now check the midpoint voltage of the vertical amplifier stage at the collector of Q504. If this voltage is less than 12 volts, suspect a shorted transistor.

Next, check for 13 volts base bias at the collector of Q502. If 0 volts, suspect shorted Q501 or Q502. Verify shorted Q502 by shorting the base of Q502 to ground. With the base shorted to ground, the collector voltage at Q502 should increase to greater than 10 volts.

Retrace Lines at Top of Picture (Scan Reduced)

For this symptom check the collector voltage of Q506. If less than 20 volts, suspect shorted Q506 and an open R520. Remember, if CR502 is open, always suspect a shorted top or bottom output transistor.

Check midpoint voltage. If greater than 20 volts, suspect a shorted retrace switch transistor Q508.

HORIZONTAL SWEEP OUTPUT CIRCUIT (SOLID-STATE) RCA CTC 99/101 CHASSIS

This RCA CTC 99/101 chassis utilizes a conventional transistor sweep output system. All chassis operating voltages are derived from the horizontal output stage. For this reason, a problem in the horizontal output stage will cause other chassis functions to be inoperable.

Note that the horizontal output transistor may fail due to over-stress caused by a fault in some other chassis component. If a horizontal output transistor is replaced, you should perform the following service checks to prevent destroying the replacement transistor in the process of locating a defective component.

Some Symptoms

- Dead set
- Shutdown occurs

Troubleshooting Procedures

Caution: Use isolation transformer and unplug ac line cord for all static checks.

Refer to simplified circuit in Fig. 3-5 along with the service procedures indicated in the numbered blocks.

Defeat the regulator circuit by shorting across SCR 100. This can be accomplished by jumping between SCR 100 heat sink and TP 12 which are not shown in this circuit.

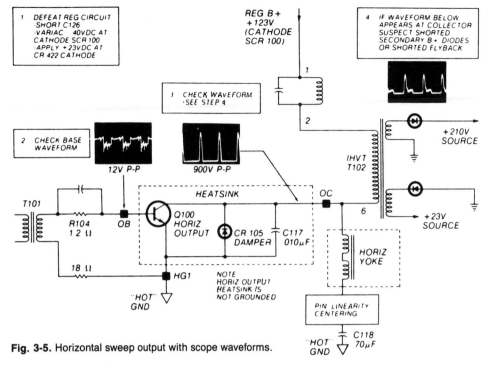

Fig. 3-5. Horizontal sweep output with scope waveforms.

Apply a reduced ac voltage through a variable transformer to supply a B+ voltage of about 40 volts. Apply +23 volts to the cathode of CR422 to operate the horizontal oscillator. Refer to Fig. 3-3.

Operating the horizontal output stage at near +40 volts allows troubleshooting of the stage while reducing the possibility of blowing out the horizontal output transistor or associated components.

NOTE: The horizontal output heat sink is *not* a hot ground. Use the negative terminal on can of C106 (B+ filter capacitor) for a hot ground.

Check the horizontal output base waveform at test point OB. If waveform is not present, suspect the horizontal oscillator driver transistor, Q407 or driver transformer, T101.

Check the horizontal output transistor collector waveform at test OC. This waveform can be very useful for problem diagnosis. If the waveform shown in step 3 of Fig. 3-5 is present and correct, increase the variac output to supply approximately +123 volt dc. Now check the secondary B+ supplies to be certain if they are not excessive. If the B+ voltages are all normal then the sweep circuit is operating properly.

If the collector waveform of the horizontal output transistor is close to that shown in step 4 of Fig. 3-5, suspect shorted secondary B+ diodes or a defective sweep transformer. To further check for a sweep transformer-short, disconnect the two secondary B+ diodes, +210 volt and +23 volt supply. If the waveform is still as shown in step 4, the horizontal output transistor is probably bad.

CHROMA CIRCUITRY (IC CHIP) RCA CTC93 CHASSIS

This sophisticated RCA U304 chroma IC contains an ACC detector, AFPC detector, chroma amps., burst keyer, color killer, 3.58 MHz oscillator, buffers and demodulators. Refer to the block diagram in Fig. 3-6 for this chroma chip circuit, and the scope waveforms that should be found for normal set operation of color reception.

Color Symptoms

- Weak color.
- Improper color.
- Improper tint.

Circuit Diagnosis

The symptoms listed above are almost always associated with a fault in the chroma circuitry. However, color faults can be caused by troubles in the video i-f circuits. Should you find no chroma input signal to the i-f circuits, then a faulty "front end" or tuner would be indicated. The tuner can be checked by inserting a video signal containing 3.58 MHz information to test point 6 located on the main chassis board. Signal amplitudes of .1 volt p-p should produce good color on the screen. Note that the color bars may or may not be in sync. Use a color-bar test generator for this check. Since all color processing operations are performed by the integrated circuit, most of the color problems will be caused by the U304 chip.

Service Checks and Tips

Use an isolation transformer and unplug ac line cord during all static test procedures.

NOTE: All waveforms are taken with the color control at maximum and a keyed color-bar rainbow input test signal. Allow the set and any test instruments to warm up 5 or 10 minutes before any tests or measurements are performed.

Points of Service

Check at pins of the chip with a scope for any missing or incorrect waveforms. Referring to Fig. 3-6, check and see if you have the correct dc voltages at all pins of the U304 IC as indicated.

With the scope, confirm that you have the correct chroma input signal at TP 6.

Check the chroma level (preset) for proper dc swing, 1 to 11 volts with a maximum to minimum color level.

Check for 3.58 MHz oscillator sinewave output signals at pins 11 and 12 of the IC.

Again, with the scope, check the keying pulse input. If the pulse is missing, there will be no color. This should be checked at the input of Q312 (station sync), and directly at pin 24 of the IC.

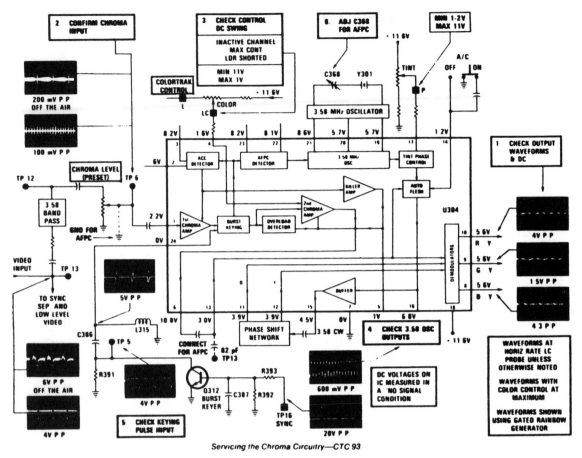

Fig. 3-6. Chroma circuit with ckeckpoints and waveforms.

Some AFPC Adjustment Tips

Adjust the 3.58 MHz oscillator trimmer capacitor C368 by using the following procedure.

- With color-bar generator, feed signal to the RF input.
- Ground TP 6. Make this ground connection as short as possible.
- Connect TP 13 (video output from i-f) to C377 (pin 13 of chroma IC) via an 82 pF capacitor.
- Adjust C368 to zero beat the color-bar pattern. Use a nonmetallic screwdriver.

HYBRID VERTICAL SWEEP CIRCUIT

This hybrid vertical sweep circuit will be found in the RCA CTC-36 and other chassis. Referring to Fig. 3-7 we see that this circuit is comprised of a transistor oscillator stage and a pentode output stage. The pentode is a special type, having a diode plate which utilizes the common cathode of the pentode. This diode, in conjunction

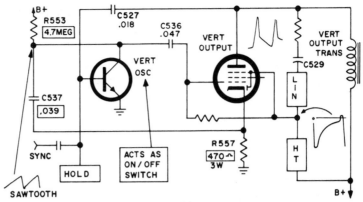

Fig. 3-7. Vertical sweep circuit with operational guide.

with a resistance-capacitance network coupled as a feedback between the output and the input electrodes of the pentode, develops a feedback voltage for addition to the sawtooth waveform for vertical linearity and size control. Another feedback path, to sustain oscillation, is from the output stage to the base circuit of the transistor oscillator. The base circuit is also the input where the vertical sync pulses are injected. The vertical control circuitry (V hold control) is designed to adjust the triggering waveform component to the base of the oscillator. The height and linearity circuits operate in conjunction with the pentode stage.

The transistor oscillator stage acts as an on/off switch. A resistance-capacitance sawtooth generator circuit (R553, C537) coupled to the grid of the output stage is subjected to alternate charging from the B+ supply and discharging through the transistor in an operating cycle recurring at the vertical sweep rate. Thus, the charging path for C537 is completed through the cathode resistor of the output tube. The oscillator transistor conducts during retrace time to provide a discharge path for the sawtooth capacitor. The sawtooth voltage developed across (C537) is coupled via (C536) to the control grid of the output stage. This grid voltage waveform, which is substantially linear, causes plate current to flow in the output stage. The plate current variation of the output stage is coupled to the deflection yoke by means of a vertical output transformer.

During the trace portion of each deflection cycle plate current increases causing plate voltage to decrease in a substantially linear manner. At the end of vertical trace time, a positive polarity synchronizing pulse is fed via a capacitor to the base of the oscillator transistor, driving the transistor into conduction. As the transistor conducts, the voltage across C537 decreases rapidly, driving the control grid of the output stage negative.

When the output stage is driven into cutoff, the plate voltage waveform increases rapidly as a result of the energy stored in the deflection yoke windings. The developed positive pulse, coupled through C527 to the base of the oscillator keeps the transistor in a conductive state. Near the end of vertical retrace, as the positive waveform decreases, the oscillator transistor is turned off. Capacitor C537 then starts to recharge in order to begin the next deflection cycle.

During retrace time the capacitor (C529) in the linearity circuit is charged through the conduction of the internal diode. The discharging of this capacitor through the linearity and height control circuits, during field scan, provides the feedback waveform for linearity. The linearity control setting determines the amount of charge that C529 receives during retrace. Thus, the proper waveform for linearity can be attained. Adjustment of the height control, however, primarily determines the average dc of the waveform, and in so doing provides the bias control for proper picture height setting.

The detailed circuit shown in Fig. 3-8 of the vertical oscillator and output stage, will now be used to point out several other features. The oscillator is provided with an additional triggering waveform component derived from one of the transformer secondary windings. The waveform is coupled to the base circuit of the oscillator via R414, the vertical hold control, and R546. Dc base bias is provided by R419 coupled from the B+ supply to the junction of the hold control and R414.

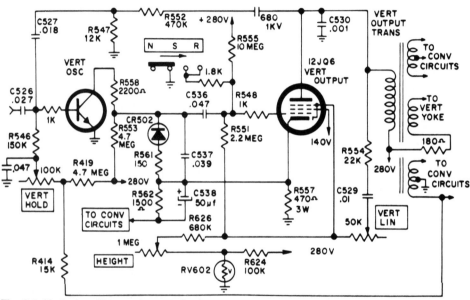

Fig. 3-8. Vertical oscillator and output circuit.

A diode (CR 502) is coupled across the sawtooth forming capacitor C537. The purpose of this diode is to prevent an instability condition, as the C537 would otherwise charge to the cathode voltage potential during retrace; any disturbance in the output tube could then be regenerative and cause some picture jitter. With CR502 in the circuit, the voltage across C537 at the start of each cycle is kept constant (due to forward drop in diode) and independent of what the cathode voltage was in the preceding cycle. Thus, by assuring that at the start of every cycle the grid drive on the tube always starts at the same point (for a given setting of the controls) the vertical jitter will be prevented.

A VDR (RV 602) is located in the height control circuit to maintain a constant reference voltage to provide height tracking with line voltage variations.

The normal-service-raster (setup) switch grounds the grid circuit of the output tube through a 1.8 k resistor when in the service position, to collapse vertical sweep for making color temperature (grey scale tracking) adjustments. The function of the 1.8 k resistor is to provide protection for the diode and transistor from picture tube arcs.

PLATE-COUPLED VERTICAL SWEEP CIRCUIT (TUBE-TYPE)

The vertical sweep oscillator turns off and on (be it tube-type or solid-state) approximately 60 times each and every second. A pulse from the sync separator stage is used to turn the vertical oscillator on at the proper time.

You can consider the height control circuit as the components between the vertical output stage and the vertical oscillator stage that control turn-on time. The vertical linearity circuit is used to connect the sweep output circuit with the vertical oscillator circuit to control this length of time. Hence, it controls the time on and off of the sweep output circuit or the actual length of time to stay turned on.

As we now see, the vertical sweep system can be broken into two basic stages. The vertical oscillator stage whose on/off timing is controlled by a sync pulse from the sync separator stage. The vertical output (amplifier) stage that is turned on and off at a 60 Hz rate by the vertical oscillator and is controlled via the height and linearity circuits. Again, the height and linearity circuits control how long the output circuit stays turned on and off. See Fig. 3-9.

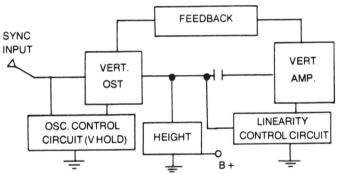

Fig. 3-9. Block diagram of plate-coupled vertical sweep system.

So, the vertical sync hulse holds the vertical oscillator to the correct speed while the vertical output stage then develops the proper raster scan size for the picture tube.

With just these two blocks the vertical circuit seems quite simple. However, things do become more complex when you consider the feedback loop from the output stage that may be used for vertical height, linearity or vertical sync pulses.

PLATE-COUPLED MULTIVIBRATOR (TWO-TUBE-TYPE VERTICAL OSCILLATOR AND AMPLIFIER)

The vertical sweep system shown in Fig. 3-10 is a plate-coupled multivibrator consisting of V1 and V2. For the oscillator portion, V2 (amplifier) is the normally *off*

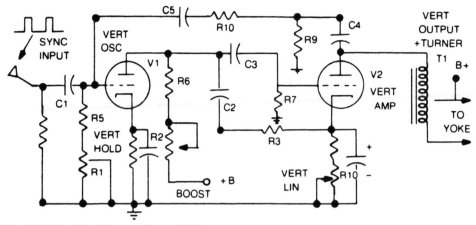

Fig. 3-10. Simplified diagram of a multivibrator.

tube. Coupling from the plate of V1 to control grid of V2 is via C3 and R7. The feedback from the plate of V2 to the control grid of V1 consists of C4, R9, R10 and C5. This network is also a shaping circuit and ensures that the voltage fed back has correct amplitude and waveform to sustain oscillation.

C1, R1 and R5 control the main frequency, determining the RC time constant of the circuit. Check these components if you have oscillator drift, which of course causes vertical picture roll. The vertical hold control R1 is located in the grid of V1, the tube that is normally off. R2 is the height control while C2 and R3 constitute the shaping network across which the output trapezoidal waveform is taken.

The output from V1 is now coupled via C3 to V2 the vertical amplifier stage. Next, the vertical output transformer couples the trapezoidal waveform to the deflection yoke. Note that a positive sync pulse is fed to the control grid of V1, the normally off tube to synchronize the oscillator.

The output amplifier found in the vertical sweep stage of a tube type TV receiver is a power amplifier. A high current carrying triode or power pentode is usually used. A power amplifier is required because the deflection yoke winding is a current driven coil. Deflection is accomplished by the changing magnetic field around the yoke, which occurs because of a changing current through the yoke windings.

The voltage waveform that appears in the control grid of the vertical amplifier, V2, has been shaped to cause linear deflection. With a scope you will see the familiar trapezoid waveform (see Fig. 3-11) which will appear in the output of the vertical amplifier.

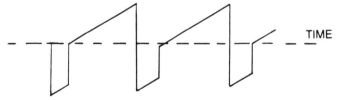

Fig. 3-11. Trapezoidal waveform for the vertical sweep.

Brief Vertical Sweep Circuit Review

A typical vertical sweep circuit (see Fig. 3-12) used in many Zenith sets features an oscillator and amplifier, with the amplifier acting as part of the oscillator. The vertical oscillator triode (V5A), discharges and couples a strong pulse into the vertical output grid, (V5B) which is a class "A" amplifier, via the setup switch and C76. This in turn couples a higher pulse to the output transformer and yoke for deflection. For the feedback action, a portion of the output pulse is looped back from the plate, pin 4, via .01 μF (C71), a 47 k resistor, then on to .0039 μF (C73) and a 100 k resistor to the oscillator grid pin 10. This triggers the next discharge pulse. The vertical sync pulse is coupled into the cathode, pin 11, at a point above the X5 diode.

Some Troubleshooting Comments

When a multivibrator stage is faulty, it is usually tough to troubleshoot the circuit while in operation unless a signal is injected into the circuit. The injected signal is then followed through the circuit (with a scope) to locate the point where the signal path is broken. The test signal can be fed in at the sync input point by using the TV stations sync pulse.

Another simple way to check the vertical sweep section is to couple an ac signal from the sets tube filament transformer winding with a capacitor over to the sync input point. This 60 Hz ac signal is then traced with a scope through the vertical oscillator stage (now acting as an amplifier) on through the coupling network and the second oscillator stage (vertical output amplifier) and then back via the signal path (feedback circuit) to the vertical oscillator section. Of course, the shapes of the waveforms will be modified by the various shaping networks. The signal amplitude and its waveform are your clues to a defect, since you will be looking for a breakdown in the signal path or a low signal amplitude which may be preventing normal oscillation.

Vertical Sweep Trouble Symptoms

The following is a list of trouble symptoms that indicate a problem in the vertical sweep system:

- Complete loss of vertical deflection.
- Partial loss of vertical deflection (may have foldover).
- Excessive vertical deflection (picture stretched out).
- Intermittent sweep.
- Vertical picture roll (may want to roll after set warm-up).
- Vertical picture jump and/or jitter (usually caused by a sync trouble).
- Intermittent or pulsating vertical sweep or roll (the bouncing ball effect).
- Vertical picture nonlinearity.

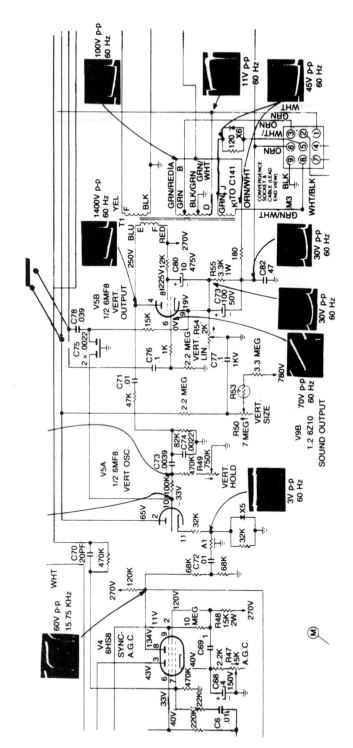

Fig. 3-12. Zenith vertical sweep tube-type circuit.

Vertical Circuit Quick Checks

Let's now look at some ways to make some vertical circuit checks to quickly isolate the faulty component. For these following circuit checks and trouble tips we will be referring to the vertical sweep circuit used in a Zenith 15Y6C15 color chassis. See Fig. 3-12.

We will begin these points of service checks with a set that has complete loss of its vertical sweep.

The first step is to open the feedback circuit at the junction of the 47 k resistor and .0039 μF capacitor (C73), which should now produce a horizontal line across the screen due to complete vertical collapse. Should complete vertical collapse not occur, the vertical output stage is oscillating itself in the absence of a vertical drive signal. Check (C82) .47 μF, (C79) 100 μF, the exact value of 15 k resistor from cathode pin 9 and the vertical linearity control (R54), as a fault with one of these components may be causing the oscillation.

Now connect one end of a .1 μF, 600 volt capacitor to the sets 6.3 volt ac filament terminal leads. Use the other end of the capacitor, through a clip lead, as a 60 Hz drive signal, to check sweep circuitry performance.

With this setup, touch the test lead to the vertical output plate, pin 4 of V5B. A one-shot negative kick will appear on the screen as the .1 μF test capacitor charges up. The vertical sweep on the sets screen should now increase to several scan lines, indicating that the output transformer circuits and the deflection yoke are good.

Should no deflection occur suspect the following components.

- Vertical output transformer or yoke could be faulty or open.
- Plate load resistor for V5A. Check for 250 volts B+ at pin 4.
- Some sets have a vertical centering pot. If open, this will cause loss of deflection.
- Check the 12 k ohm screen grid resistor and capacitor C80.
- Vertical linearity control R54 may be open.
- Also faults in the pincushion circuit and on the convergence board will cause loss (sometimes partial) of the vertical sweep.

Partial Vertical Deflection Checks

Some shorted turns of the vertical output transformer can cause partial loss of vertical deflection. A more common trouble is an open capacitor, C73, from cathode to ground of the V5A tube. For a quick check just bridge a 100 μF, 450 Vdc capacitor across C73. Make sure you connect the test capacitor with the proper polarity. Note that on some brands of sets this capacitor may be located on the convergence board. Also, R55, the 3.3 k ohm cathode resistor may have increased in value. Just bridge another resistor across to see if you now have full deflection sweep. The vertical sweep output cathode circuit is also returned through some of the controls on the convergence board. So, check these, they may be open or you may find a crack in the PC board.

If the sweep circuit has checked out OK up to this point, then inject the test signal into the vertical output tubes control grid, pin 6. Since this stage is a class A amplifier, the scan on the picture tube should display about 3 inches of vertical sweep (sinewave, not sawtooth) which will indicate that the output stage is good.

With the 60 Hz test signal fed to the vertical oscillator plate, pin 2 of V5A, an identical deflection to that observed on the vertical output grid, indicates that the setup switch and the (C76) coupling capacitor are good. If no deflection is observed then suspect a faulty (usually open) coupling capacitor between plate of oscillator and control grid of the vertical output tube. Also suspect any capacitors tied to the coupling network (such as C75) as they could be shorted or have some leakage and will bleed off the signal before it reaches the output stage grid. Also, look for open or increased-in value resistors in this network. Also, a faulty setup switch S1, will cause a loss of vertical deflection or erratic vertical sweep action in this circuit. The switch may have developed leakage to ground or the contacts may be dirty and this will cause the vertical sweep deflection to be intermittent.

Oscillator Grid Injection Checks

Before placing the test capacitor on the vertical oscillator grid, pin 10, connect the set for normal viewing by tuning in an active TV channel. Now clipping the test capacitor to pin 10, should create a full picture scan, indicating the vertical circuit is functioning properly, except for the disconnected feedback loop. Picture will tend to roll around vertically (see Fig. 3-13) since the triggering (sync-lock) is at the 60 Hz ac line frequency rate.

Fig. 3-13. Picture roll caused by 60 Hz ac in the oscillator circuit.

Connecting the test capacitor to the loose end of the .0039 μF capacitor (C73) in the feedback circuit should produce a vertical picture sweep, but at a reduced amplitude. You can also hold the (C73) capacitor between your fingers and obtain some vertical sweep, amplitude being dependent upon the amount of random 60 Hz ac picked up via your body capacitance.

With the feedback circuit now connected, you should have normal vertical sweep. If not, then some component in the feedback loop (between plate of output stage and control grid of oscillator) is open, or a bypass capacitor to ground, or a capacitor to another portion of the vertical circuit is shorted. The prime suspects would be C73, C71, C74, and the 100 k and 47 k resistors.

In order to keep oscillating, the vertical circuit relies on the 60 Hz pulse that comes from the plate of the vertical output stage, through the feedback loop, and to the control grid of the oscillator stage. Without this pulse, the vertical oscillator stage would actually go into saturation, thus shutting down the whole vertical sweep operation.

In summation, when an oscillator (multivibrator) is faulty, it can be tough to troubleshoot when it is in operation unless a test signal is injected into the circuit. This test signal can then be followed with the scope in order to pinpoint the place of signal path interruption. If the TV station sync pulse cannot be used, then a test capacitor is an easy way to couple a 60 Hz ac signal from the sets tube filaments to the vertical circuits sync input point. The test signal is then traced with a scope through the oscillator stage (now acting as an amplifier) through the coupling networks and onto the vertical output stage and then through the feedback circuit to the first stage again.

Vertical Oscillator Frequency Checks

Loss of vertical frequency control (no picture lock or vertical roll) may be caused by loss or attenuation of the vertical sync pulses, or to value changes in the frequency determining components in the oscillator stage.

You can easily determine if the oscillator is at fault by varying the vertical hold control. If the picture can be rolled up and down, it is almost a sure bet that the oscillator is operating properly and the trouble is caused by loss of, or improper sync pulses. You should now suspect a fault in the sync circuits or in other circuits that supply information to the sync stages.

If rotation of the vertical hold control will not roll the picture in both directions, then look for trouble in the oscillator circuit. Probably the quickest way to locate the fault is to check out or sub-in each component in the oscillator circuit as there are only a few. Prime suspects would be the tube itself V5, or C73, C74, 100 k resistor, 470 k resistor, C75 special dual-capacitor and the vertical hold control, R49. The fast vertical roll in Fig. 3-14 was caused by a leaky C73, .0039 μF capacitor.

If diode X5 in cathode of V5A becomes shorted there will be no vertical picture lock-in. This fault causes the sync pulses to be shunted to ground. An open X5 diode and/or value change of the 32 k resistor will cause no vertical sweep or a reduced

Fig. 3-14. Leaky capacitor causes fast vertical roll and horizontal bending.

deflection. A pulsating vertical sweep, sometimes referred to as the bouncing ball effect, is caused by a faulty A1 integrator RC network.

Vertical picture jitter and jump is usually caused by components that are intermittently going bad. Also, a hum bar creeping up the screen will make the picture jitter. A light shade drifting bar is usually due to a signal path trouble, while a black bar is a fault in the dc power supply, such as an open filter capacitor. And do not overlook leakage between multi-section filter capacitors.

Vertical Picture Foldover and Stretch

In most cases, the reason for picture foldover, is improper bias on the output amplifier tube, which results in overloading, or an improper sweep-voltage waveform fed to the output tube.

If the trouble is caused by wrong bias, a scope check at the grid, pin 6 of V5B, will probably show a clipped waveform at this point. Our job then, is to find the reason for this clipping.

A leaky coupling capacitor C76, a shorted cathode bypass C79, or a shorted vertical linearity control R54, would upset the bias on the vertical output stage and cause foldover.

A voltage check at the control grid should give you a clue to the problem. Thus, a positive voltage metered at the control grid of V5B, would indicate a leaky capacitor C76. And of course, a zero cathode voltage reading would point to a short in the cathode circuit.

A long stretched out picture is generally caused by too much vertical drive signal from the vertical oscillator. The stretched out picture shown in Fig. 3-15 was caused by a faulty R54 VDR in the height control circuit. The clue, in this case, was way too high plate voltage, pin 2 of V5A, on the vertical oscillator stage.

Fig. 3-15. Faulty VDR causes picture to stretchout.

Nonlinearity of Vertical Sweep

Improper operation of the vertical amplifier, which causes a nonlinear picture as shown in Fig. 3-16, is usually caused by improper grid bias or low voltages on the plate or screen grid. A defective linearity control may cause the picture to be nonlinear. And a leaky coupling capacitor C71 can also cause nonlinearity without foldover. In both cases the amplifier is affected. A waveform, voltage and resistance analysis of the V5B amplifier should point you in the right direction. Leakage in C80, an increase in value of the 12 k resistor at the screen grid, pin 8, or a defective T1 vertical output transformer can cause a nonlinear sweep. And of course, open filters in the B+ voltage supply can cause the same symptoms. You may or may not see a hum bar across the screen.

Nonlinearity may also be caused by a nonlinear sweep waveform coming from the oscillator stage. Thus, if the values of C77 or the 2.2 meg resistor have changed, the waveform from the oscillator could be distorted. Waveform, voltage, resistance checks or parts substitution in the oscillator stage should correct the trouble.

THE ZENITH SYNC/AGC PROCESSOR CHIP 221-45

Because of tremendous TV signal variations and interference found in actual reception, some complex IC circuitry has been designed in order to produce a stable picture. This, of course, is referred to as the Sync/AGC circuitry. In the Zenith sets this sophis-

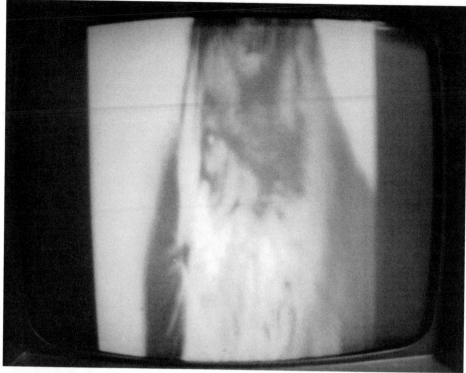

Fig. 3-16. Nonlinear vertical sweep and reduced width.

ticated chip is known as the Sync/AGC processor. It is found in the 9-87 plug-in module.

Briefly, the AGC circuit senses the amount of signal received by the TV set after it has been amplified by the tuner and i-f stages. A dc control voltage developed by the signal strength is then fed back to the tuner and i-f amplifiers to adjust for proper output signals. In order to have a good signal-to-noise ratio, the gain of the i-f amplifiers is reduced before the tuner gain is reduced.

To help you see the total picture for this processor chip operation, refer to the block diagram in Fig. 3-17 of the complete Sync/AGC system which is contained on the 9-87 plug-in module.

The video information that the TV station transmitter broadcasts is, of course, amplitude modulated. This AM video signal will vary according to picture information and cannot be used as a reference to produce the AGC voltage. Because the horizontal sync pulse does not vary, except for changes in signal strength, it is compared to an internal reference to develop the AGC correction voltage.

The AGC circuitry uses the horizontal sync pulses which are removed from the composite video signal by the sync separator circuit. They are then processed through a dual-time constant network to produce both horizontal and vertical sync pulses that are used to lock in the vertical and horizontal oscillators. The complete video signal enters the processor chip and the sync separation function is performed from within.

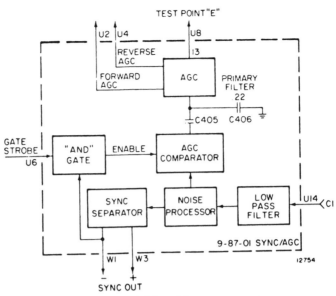

Fig. 3-17. Functional block diagram of the 9-87 module. (Courtesy of Zenith Corp.)

Now we will take a close look at the Zenith processor IC, how to troubleshoot around it and its associated components plus, review some actual case history problems.

Video Processor Circuitry Operation (Zenith 9-87 Module)

The heart of the 9-87 plug-in module is the 221-45 video processor chip. Two signals are fed into the IC401 chip. One is the composite signal from the i-f module and the other is the gating pulse from the horizontal circuit. The IC processor uses these two inputs to develop AGC voltage and sync information as well as performing noise gating and sync-limiting. As we look at the various processor operations refer to the 9-87 module circuits in Fig. 3-18A. The correct scope waveforms (input and output signals) will appear as shown in Figs. 3-18B through 3-18E when all receiver systems are functioning.

Negative going composite video at test point C1, goes via a high frequency filter network formed by choke coil L211 and capacitor "C220 to pin U14 on the 9-87 module. The function of this filter network is to remove high frequency components on the composite video signal, which could in poor reception areas interfere with normal sync limiting action. These filter components could be prime suspects should you encounter poor sync locking action and a replacement module does not solve the problem.

The composite video signal enters the IC at pin 8 via a 1 k resistor. The chip at this point performs the noise processor function. When noise pulses beyond sync tip level appear at pin 8, IC 401 cuts off the AGC and sync functions for the duration of the noise pulse. Turning off the sync function prevents the noise pulses from tripping the vertical and horizontal oscillators.

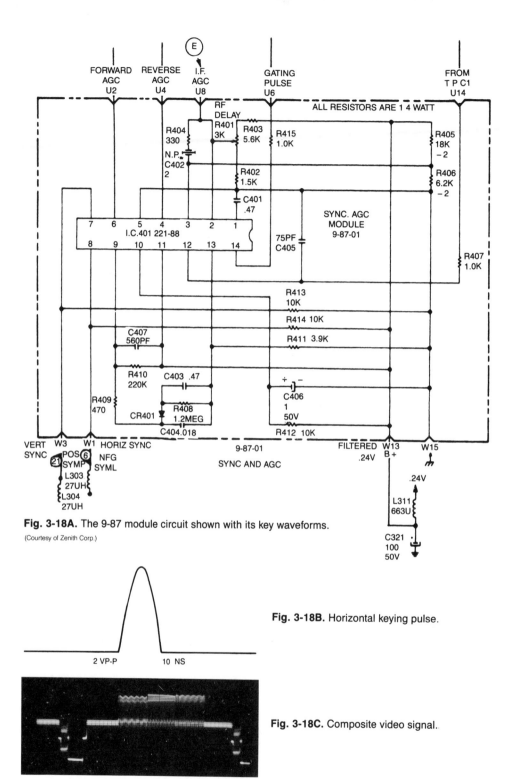

Fig. 3-18A. The 9-87 module circuit shown with its key waveforms.

(Courtesy of Zenith Corp.)

Fig. 3-18B. Horizontal keying pulse.

2 VP-P   10 NS

Fig. 3-18C. Composite video signal.

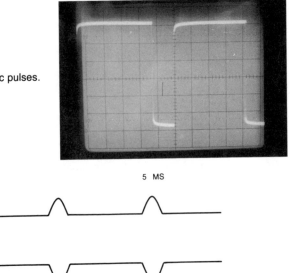

Fig. 3-18D. Horizontal sync pulses.

10V P-P   5 MS

Fig. 3-18E. Vertical sync information.

Noise Protection and AGC Gate Circuit (221-45)

Within the sophisticated 221-45 chip, is an AGC comparator which is gated so that fast slew rates can be utilized with good noise immunity. You will note in Fig. 3-19, the noise protection block diagram, that two gates are used. In order for the AGC filter to be charged or discharged, two inputs to the comparator must be present. As with other

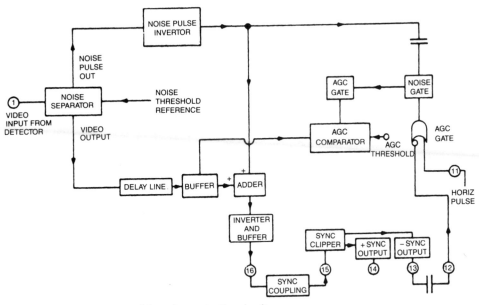

Fig. 3-19. Block diagram of the noise protection circuit.

discrete AGC systems, the keying pulse must be present but additionally, the negative going separated sync pulse from the video signal must be present. Coincidence of the two pulses will result in AGC action if the detected signal level changes. If, however, the sweep oscillator is out of step, no AGC action can occur until the keying pulse coincides with the signal sync pulse. Thus, no sampling or wrong sampling is made during scan time.

AND Gate Note

As Fig. 3-19 indicates there is a NOT input to the AND AGC gate in the noise protection circuit. Note the NOT circle at one of the gate inputs. Remember that the NOT symbol indicates logic inversion. Just a note in passing, that you will now start finding these hybrid NOT-input gates in some of the newer generation VCR logic system circuits.

As the noise cancelled video signal leaves the IC at pin 5 it enters a dual time constant network of C403, C404 and diode CR401. This network provides the sync limiting stage with immunity to aircraft signal flutter.

After passing through the dual time constant network, the noise cancelled video signal enters IC401 at pin 4 and the sync limiting function is performed. Note sync separation location in Fig. 3-19. Internally, IC401 separates the video information from the sync and amplifies only these pulses. A positive going sync pulse appears at pin 3 of the IC and is the vertical sync. At pin 2 a negative going pulse is used to sync in the horizontal sweep oscillator.

The negative-going sync information at pin 2 is returned to pin 1 thru R404 and C402. The sync pulse is coupled to a gating circuit within the chip. Also, a positive-going horizontal gating pulse is coupled from the horizontal circuit thru R401 to pin 16 of the IC. Both the negative-going sync pulse and gating pulse is needed to key *on* the gating state. AGC voltage can only be developed when sync is present which is during the retrace interval. This action occurs in Fig. 3-19 along the enable line from the AND gate to the AGC comparator stage.

Since AGC voltage can be developed only when sync is present, AGC voltage must be maintained during scan time. Capacitor C406 at pins 10 and 11 of the IC functions to hold this voltage. Also, if noise is present during the sync interval, the sync output will be cut off and C406 will hold the AGC voltage until the next gating pulse.

With no signal present, maximum i-f AGC gain is set by the voltage divider comprised of R410 and R418. This divider sets close to +6 volts at pin 12, which in turn sets up a bias on a portion of IC401 that results in about +4.8 volts at pin 13. The +4.8 volts is added to the +dc voltage set by the AGC delay control. The resultant voltage is coupled thru R406 to the i-f AGC output voltage at pin U8 of the module. On some sets this is flagged as test point *E*. When a strong TV station is tuned in, the internal circuitry of the IC increases the output voltage at pin 13 until maximum gain reduction occurs. This voltage is about +7 volts and is determined by the adjustment of the AGC delay control.

In addition to the i-f AGC voltage, the IC produces two RF AGC voltages. The AGC voltage at pin 14 is for the field effect transistor (FET) RF stage in the VHF tuner. The AGC voltage present at pin 15 is used by the NPN bipolar transistor RF stage in the UHF tuner. These dc voltages on pin 14 and 15 will vary according to signal strength and type of tuners used in the set. You will not find an AGC level or gain control on these chassis that have the 221-45IC. This is because of the IC processor design and tight component tolerances in the 9-87 module.

Troubleshooting the Processor Circuits

In this section we will not dwell on repairing the 9-87 module (a new or rebuilt one can easily be plugged in) but rather in checking the modules associated components, voltages and pulse waveforms. As we now look at these various troubleshooting tips and case history problems, refer back to the 9-87 module circuit in Fig. 3-18A along with the key scope waveform checkpoints for reference information.

When you have a symptom that appears to be a Sync/AGC problem, some circuit checks and observations must first be performed in order to get a handle on the situation. The first step is to determine if the problem is actually caused by a fault in the AGC and/or sync circuitry. Symptoms of weak video, loss of video, or distorted picture as well as loss of vertical and horizontal sync can be caused by faults in the tuner, i-f stages, Sync/AGC stages or power supply. The best way to start checking these sets to determine if the Sync/AGC stage is defective is to sub-in a good module for the one in question, and then adjust the AGC delay control. If this does not correct the problem, then go to the tuner and i-f circuits for video problems and the vertical and horizontal circuits for sync problems. Also, check for +24 dc at pin W13 and correct horizontal keying pulses at pin U6 of the 9-87 module. Should you not have a good substitute module, a variable dc power supply can be used to clamp the tuner and i-f amplifier AGC voltages to normal operating levels. If a nearly normal picture cannot be restored, the defect may very well be in the tuner, i-f stages or video circuits. At this point a tuner "subber" will help you pin-point the fault quickly.

Let's now look at some "real world" problems with this Sync/AGC module and IC.

The Picture Symptom

In this Zenith 19FC45 color chassis the picture would roll intermittently and tend to twist. On some channels several wide bands or streaks (see Fig. 3-20) would go across the screen horizontally. This picture trouble may or may not be of an intermittent nature. The top of the raster may also be of a lighter shade.

The Cause and Solution

The problem area, in this case, was found to be around the 9-87 Sync/AGC modules associated circuits. To pin down this problem several scope checks were made at various +24 volt input module pins around the chassis. When the symptom occurred

Fig. 3-20. Streaks caused by open filter capacitor at +24 volt input at pin W13.

the scope indicated a lot of hash at pin W13, +24 volt input terminal of the 9-87 module. The hash (horizontal pulses from HV sweep stage) was due to loss of filtering because of an open C321, 100 μF bypass filter capacitor. A replacement filter cleared up this picture problem.

For normal operation at all points where the +24 regulated voltage enters the various modules the scope trace should be a smooth line even at high vertical scope amplifier gain. If you find hash or spikes on the scope look for open filter and bypass capacitors or trouble in the 24 volt regulator circuits. Such as a faulty zener diode or regulator transistor. Also, make sure that the +24 volt line measures 24 volts dc. A problem in the regulator circuit can lower this a few volts and cause a picture symptom that looks like sync or AGC trouble.

No Sound or Picture—Blank White Screen

This is a very common and simple problem but can be overlooked very easily. If module replacements do not restore operation check for +24 volts at pin W13 of the 9-87 module. If no voltage is found, check for an open LX311 filter choke coil in the 24 volt line. This is a small 663 μH coil that may open up or at times will become intermittently open. Loss of voltage could also be caused by a shorted C321 filter bypass capacitor.

Dual-Trace Scope Phase Comparison Check

A rare but tough to locate AGC problem can be caused by a phase shift in the horizontal keying pulse. An easy way to locate this type of trouble is to use the dual-trace scope for simultaneously viewing the AGC keyer pulse (strobe) and horizontal sync pulse for correct timing. This will allow you to make an exact analysis of the two superimposed pulses. To check for this phase shift problem connect the synced in channel of the scope (usually the A channel) to pin 14 of the 9-87 module (video signal) and the B scope channel to pin U6, the horizontal keying pulse. Set scope at the horizontal sweep rate and adjust scope sync controls for a solid lock-in of the two pulses. The dual-trace scope photo in Fig. 3-21 illustrates the correct timing for a normally operating set. The horizontal sync pulse and strobe pulse must line up precisely.

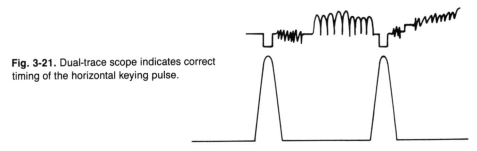

Fig. 3-21. Dual-trace scope indicates correct timing of the horizontal keying pulse.

A mistimed keying pulse can be caused by some defect in the horizontal phase detector stage or any pulse coupling components. The dual-trace scope waveforms (Fig. 3-22) readily illustrate this out-of-time condition. Note that the keying pulse is actually in time with the 3.58 MHz color burst signal.

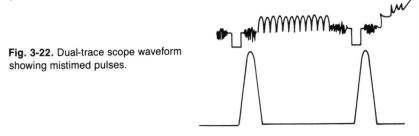

Fig. 3-22. Dual-trace scope waveform showing mistimed pulses.

Another good point is to take note of the keying pulse width. If the pulse is too wide some video or 3.58 MHz color burst information may be sampled instead of the horizontal sync pulse level and this could cause the wrong AGC dc voltage to be developed. The symptoms would be picture overload, perhaps some bending or a washed out picture.

Another point of service is to check for proper filtering of the dc AGC control line. If a filter capacitor opens on this line the scope will show some hash. An open AGC line filter will cause various AGC picture problems such as overloading, bending and streaks across the screen.

Mysterious Picture Bending, Weaving and Rolling Symptoms

This intermittent picture problem (it appeared as an unstable sync or AGC fault) would occur on the local cable TV system in our area only on channels that were picked up via the satellite system such as Home Box Office, etc. This picture weaving symptom (see Fig. 3-23) would only occur in Zenith color sets using the Sync/AGC modules with the 221-45 video processor chip. A new or rebuilt module has always corrected the problem. However, just replacement of the 221-45 chip will not always eliminate all unstable pictures on these cable channels.

I have found some very fine and faint noise pulses that were detected on a very wide-band scope that appeared to be mixed with the horizontal sync pulses. Could this be some type of noise from outer space? If any reader has some ideas about this phenomenon, please let me know.

Perhaps these fine pulses were activating the noise protection circuit in the chip (due to one or more off-value components on the module and/or within the chip) and caused the IC's logic circuits to "gate-out" most of the sync pulses.

Fig. 3-23. Picture weave and roll symptoms found on local cable channels received via satellite.

Some Module Checks and Tips

In about 90% of the cases a new 221-45IC will solve any trouble that occurs with this module. As the circuit indicates, this module does not contain very many discrete components.

While making a quick visual check of this module be on the alert for burnt or discolored resistors. Then use an ohmmeter to check for any off value resistors. You may then opt to check out the capacitors and the one diode.

78

For any intermittent problems on this module look for poor solder connections and cracks in the PC runs. Always clean and tighten all of the terminal pin spring contacts before the module is reinstalled. Also, check and clean pins on the main chassis where the module plugs in.

VIDEO AMPLIFIER CIRCUIT (SOLID-STATE)

This low-level video circuit is found in the RCA CTC 93 and other similar RCA color TV chassis. One circuit that may be new to you is the differential amplifier.

Video Circuit Operation (Part A)

Composite video signals from the i-f are fed to a common 3.58 MHz trap and then go via a delay line (DL 300) which feeds video to the base of differential transistor "A" (Q303). Refer to video circuit diagram, Fig. 3-24. Transistors Q303 and Q304 form a differential pair; while Q302 is the constant current source for the emitter-coupled pair. The more current supplied by Q302 to the differential pair Q303-Q304 will mean more contrast or gain.

The constant current source is controlled by the contrast control setting, LDR (Light Dependent Resistor) and action of the beam limiter via contrast buffer transistor Q301 (controls dc voltage supplied to the base of Q302 constant current source). At this point in the circuit, not only does the dc control voltage affect contrast, but also the color level.

After video signal contrast level has been set at collector of Q304, the signal is then coupled to the base of Q317 for video peaking. This stage provides stabilizing feedback voltage back to the base differential transistor Q304. The first video, Q318, provides for some additional gain and its emitter has a contrast preset control that is factory set.

Low-Level Video Circuit Operation (Part B)

The second video amplifier couples video (via C362) to the video buffer Q306. See the video circuit in Fig. 3-25. With reference to waveform at emitter of Q306, sync tips have been removed and dc is restored at base of Q306 by action of the clamp stage Q307. Positive horizontal pulses coupled to the emitter of Q307 (via R369 and CR302) cause conduction of this stage, clamping capacitor C362 to a reference dc voltage. This reference voltage is set via the brightness control and goes to the base of Q307. If the horizontal pulses are missing, the clamp stage will not conduct.

Horizontal and vertical blanking pulses are elevated by action of the blanker stage, Q310. Both vertical and horizontal pulses are supplied to and amplified by Q310 and matrixed with the composite video signal on the base of Q309. In the waveform on the base of Q309, the elevation in blanking level at this point should be sufficient to drive CRT to cutoff during horizontal and vertical retrace. Diode CR303 carries the bias current due to R371 when blanker is off. Diode CR311 provides protection for blanker transistor Q310 to prevent overstress of the transistor when blanking pulses are applied.

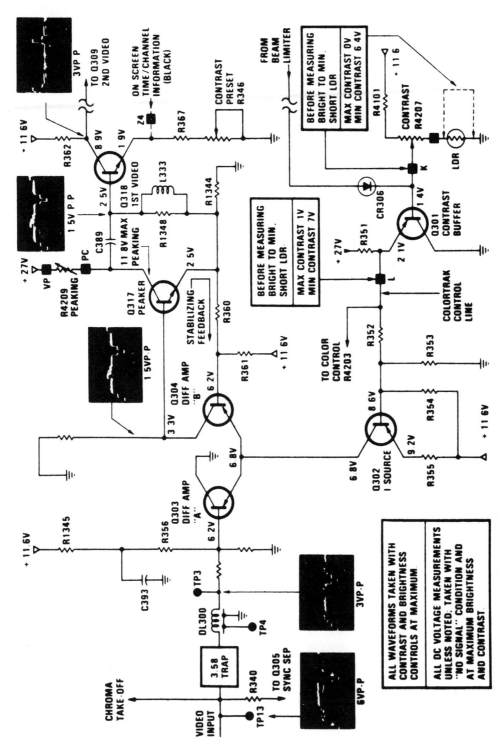

Fig. 3-24. Low-level video stage with checkpoints.

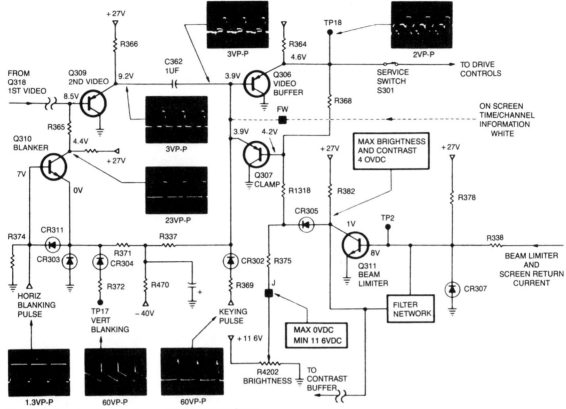

Fig. 3-25. Low-level video stage with scope waveforms.

When the clamp stage is operating properly, the video signal black level is referenced to the horizontal retrace blanking level and dc reference is effectively restored. The brightness control (R4202), as well as the beam limiter (Q311), can alter the base reference voltage of Q307, altering the reestablished "black" level reference.

The video signal is then coupled from the emitter of Q306 via the service switch and respective drive controls in the R, G, and B matrix-amplifier stages.

Basically, beam current is caused to flow to the B+ supply through R378 (reference current is established) with a portion diverted to ground through the base emitter junction of beam limiter, Q311. Transistor Q311 is normally operated in a saturated state under normal brightness and contrast settings. As beam current increases and exceeds reference current, collector voltage of Q311 will increase and cause the brightness to decrease via CR305 and contrast to decrease via CR306. Thus, beam current is limited by a combination of reducing the video drive and brightness. A filter network removes both vertical and horizontal components found in the beam limiter current applied to Q311. Diode CR312 and R332 make up a beam limiter peak overload detector, allowing fast response to large beam current overloads. If too much beam current appears at Q311, CR307 acts as a protection device to protect Q311 from breaking down.

Video Servicing Procedures

Let's now look at some low-level video circuit troubles and techniques for trouble-shooting them.

Video Trouble Symptoms and Diagnosis

- No or distorted video
- No or improper contrast control
- No or improper brightness control
- No control or contrast/brightness/color

The low-level video processing circuitry has several direct coupled stages. Therefore, it is essential that the procedure described below be followed when troubleshooting any of the above symptoms.

Note that for sets equipped with FS Remote Scan Tuning Systems, there is essentially a built-in troubleshooting aid with the on-screen time/channel display. If no raster appears, but time and channel information appears on the screen, you can then assume that the low-level video stages signal path is OK from Q318 (1st video) to output of CRT.

Troubleshooting Procedures

Use isolation transformer and unplug ac line cord during all static checks. Refer to the block diagram (Fig. 3-26) for these procedures.

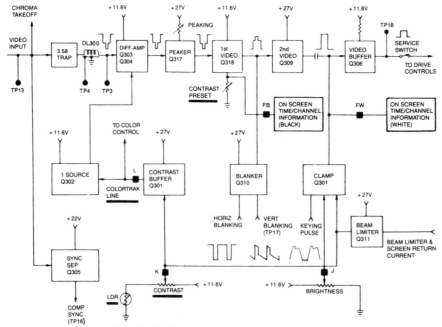

Fig. 3-26. Block diagram of low-level video stage.

No or Distorted Video

Check setup switch. An open setup switch will cause a loss of raster or a very dim one as well as a loss of video.

With a scope confirm that you have a video input signal of about 6 volts p-p at TP 13. Now check the dc bias and circuit operating voltages, as well as waveforms at various video processing stages.

Check dc voltage on collector of Q311 beam limiter transistor. If dc voltage is between +18 and +27 volts, suspect open Q311 beam limiter transistor. An open transistor at this point makes both contrast and brightness controls inoperative. Screen will be totally black, even if chroma is turned to maximum.

Check blanker stage Q310, as improper operation can result in insufficient brightness.

No or Improper Contrast Control

Under this condition, screen will normally appear washed out and chroma control will have little effect. Check various dc voltages associated with the contrast control (test point K) and contrast buffer (Q301). Also check the color track control voltage (test point L), constant current source (Q302), and differential transistor pair Q303 and Q304. If dc voltage at test point (L) reads about +11 volts, assume contrast buffer Q301 may be open. Check diode CR306. If this diode is shorted the contrast control will not operate.

No or Improper Brightness Control

Check Q307 clamp action by scoping the base and collector of Q307 while varying the brightness control. As brightness control is decreased (ccw), video should go in the direction of black.

Confirm proper dc voltage range of brightness control (R4202) at test point (J). To isolate beam limiter stage from the brightness control and clamp stage, ground collector of Q311. If problem clears up, fault is most likely open beam limiter transistor Q311. If CR305 is shorted, this will cause the brightness control to be inoperative.

No Contrast/Brightness/Color Control

The very first check should be the collector voltage on Q311. As stated earlier, an open transistor in this circuit will cause all video and chroma stages to be inoperative. To remove any Q311 influence on video, contrast, and chroma, ground the collector of Q311.

POWER SUPPLY CIRCUIT
(FULL-WAVE VOLTAGE DOUBLER, S-S DIODES)

This power supply circuit is used with TV sets that do not have a power transformer. Refer to Fig. 3-27 for this circuit as used in an RCA CTC 22 color chassis.

Circuit Operation

A 7 Amp. fuse in series with one side of the ac line is used for circuit protection. Surge protection is provided by a thermistor in series with the B+ rectifiers. The B+ circuit is protected by a special thermal reset circuit breaker. This circuit breaker, CB101, can be tripped by a combination of excessive current demanded of the low voltage power supply and/or excessive current via the horizontal output tube cathode circuit.

The CB101 circuit breaker has a resistive winding (about 1.3 ohms) that completes the cathode ground return from the horizontal output tube. If cathode current becomes excessive, the winding heats, causing the thermal bi-metal strip to expand unequally; the metal flexes, causing the breaker switch contacts to open, breaking the B+ line. The same action occurs if the B+ current increases due to a short circuit or overload. Degaussing is initiated by the customer, using a spring loaded switch located on the back of the receiver.

The power supply drawing shown in Fig. 3-27 is actually a full-wave voltage doubler circuit. You will note that capacitor C106A functions differently than in a conventional power supply, as it does not act as a filtering capacitor. Operating with diode CR102, C106A charges to the peak value of the applied ac voltage. Using this dc charge as a reference, the other diode rectifier CR101 produces a dc voltage approximately equal to the peak-to-peak value of the ac line voltage supply. Hence, the dc voltage obtained is approximately two times (thus almost doubled) the rms value of the input ac line voltage.

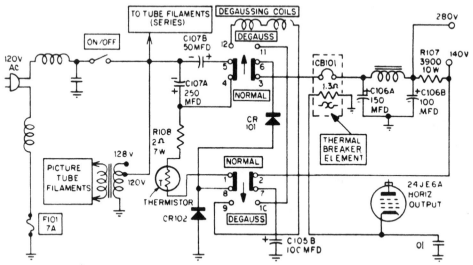

Fig. 3-27. Power supply and degauss circuit.

Power Supply for RCA CTC 42X Chassis

This power supply, used in the RCA CTC42X chassis shown in Fig. 3-28 has two separate power supply systems. The first is a full-wave doubler furnishing 280 volts B+

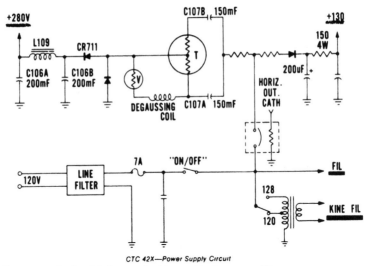

CTC 42X—Power Supply Circuit

Fig. 3-28. Power supply circuit-full-wave doubler and half-wave rectifiers.

to the deflection, chroma, video output and audio output circuits. This power supply has provisions for automatic degaussing, using the circuit that was introduced in the RCA CTC22 chassis (tapped thermistor and split charging capacitor). The other power supply is a half-wave rectifier furnishing +130 volts to the solid-state video i-f, low-level video, Sync/AGC circuits, and various tube screen grid voltages.

Power Supply Circuit Points of Service and Tips

The following is a list of various power supply circuit problems.

- No dc voltage
- Low dc voltage output(s)
- Ac ripple that may cause bar to drift up or down the screen
- Picture will wave or weave
- Circuit breaker trips off
- Shorted diodes and filter capacitors

Refer to the circuit in Fig. 3-27 as we look at these various points of service and other information.

Complete loss of dc voltage could be due to a faulty on/off switch, CB101 circuit breaker, capacitors C107A and/or B, thermistor or degauss/normal switch.

Loss of the 140 volt dc could be caused by an open R107 and/or shorted filter capacitor.

Low dc voltages will usually be caused by faulty filter capacitors (open or leaky) and diode rectifiers (open). All sides of the picture will be pulled in.

A circuit breaker that trips-out immediately is usually caused by a shorted diode CR102 or CR101.

85

A breaker that trips after about 30 seconds is usually caused by a problem in the horizontal output stage.

A set that operates an hour or more before the breaker trips is usually due to a faulty (weak) circuit breaker.

If the sides of the raster pull in and weave, suspect filter capacitors C106A and/or C105B to be open. You may also see a bar that drifts through the picture.

If both sides of the raster pull in, suspect an open filter capacitor or a weak horizontal output tube. If the right side pulls in, then look for troubles in the horizontal sweep stage.

When checking for a faulty diode rectifier use the lowest range of your ohmmeter and measure it *in circuit* as follows.

Diode Quick Checks

With the ohmmeter set to low range (X1) take a front to back ratio measurement (read in one direction, change polarity and make another reading). If the diode is shorted it will read dead-short for both polarities. If the diode is open, the ohms value (resistance of in-circuit components) will read the same for either polarity check. Should the diode be good, you will now measure a lower resistance with one polarity than with the opposite polarity.

HORIZONTAL SWEEP-TRACE/RETRACE CIRCUIT (RCA TYPE)

The following horizontal sweep (trace/retrace) circuits will be found in RCA CTC72 and CTC81 color TV chassis and are similar to those found in many XL100 sets.

Oscillator Circuit Operation

The horizontal AFC and oscillator circuits of the CTC74 are contained in the MCH module. The AFC stages, along with the horizontal blocking oscillator and 33-volt zener diode, are all located on this module.

The block diagram for the MCH-001 module is illustrated in Fig. 3-29. These stages include the Q1 phase splitter device, whose output is coupled to a pair of diodes for horizontal AFC, and thus control the horizontal blocking oscillator. Positive retrace pulses from the sweep transformer are first shaped by the sawtooth stage before being fed to the AFC stage. Operating B+ for the module stages is derived from a 33-volt zener diode, initially fed from the 175-volt source via the dropping resistor.

A horizontal range control is connected in series with the customer horizontal hold control. Readjustment of the horizontal range control is seldom required, even when another module is installed.

The horizontal oscillator is a blocking type, connected as in many other XL100 RCA chassis. The output of the oscillator stage is used to initiate retrace, by "triggering" the retrace ITR into conduction.

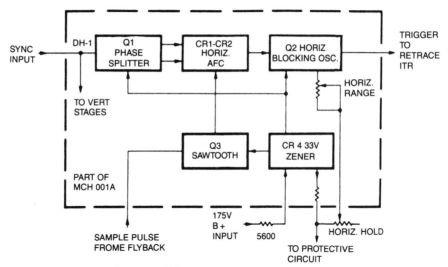

Fig. 3-29. Block diagram of horizontal IC stages.

Horizontal Trace/Retrace Circuit Operation

Figure 3-30 is a diagram of the trace and retrace circuits as used in the CTC 74 chassis and is similar to other XL-100 chassis.

Intrinsic thyristor rectifiers (ITR 401 and ITR 402) function in circuits similar to those using discrete SCR and diode pairs. ITR's were first used in the CTC 62 chassis, but these trace and retrace circuits are now more simplified.

Voltages in the CTC 74 chassis are developed directly from the power transformer. Thus, T402 no longer has secondary power windings for these voltages and the retrace circuit is more conventional in design than previous SCR systems.

The CTC 74 deflection circuits use a high-voltage transformer with an isolated primary. The top and bottom pin correction transformer is connected in parallel with the primary; the linearity network is in series with the yoke, and both are in shunt with the primary of the sweep transformer.

Operating power (175 volts) is derived from the main power supply circuit. This 175 volts (applied to the horizontal circuit) is obtained via a plug interlock connection and a relay in the power supply. The B+ path is completed by a jumper connection on the yoke plug. For this reason the yoke has to be connected before B+ is fed to the horizontal stages. In the CTC 74, ITR 401 functions as the trace SCR diode combination and ITR 402 as a dual retrace device. The gate signal from the horizontal oscillator to the gate of the retrace ITR is similar to other RCA chassis. Note that two .027 microfarad commutating capacitors are connected in parallel. Key troubleshooting waveforms would be the trigger pulse at the gate of ITR 402, the commutating waveform at the anode of the device, and the pulse waveform viewed at the high side of the trace ITR. Also important is the gating waveform at the gate of trace device ITR 401. A keying pulse for the AGC circuitry is obtained via the 150 picofarad capacitor connected to the top side of the sweep transformer primary winding.

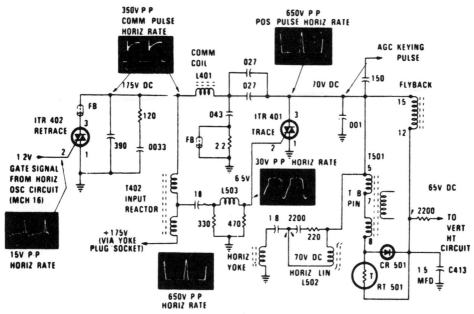

Fig. 3-30. RCA trace and retrace horizontal sweep circuits.

Because this chassis utilizes a self-regulating power input transformer, there is no need for a high voltage regulating circuit—either in the CTC 74 or CTC 81 chassis.

The simplified schematic in Fig. 3-31 shows the trace-retrace circuit for the CTC 81 chassis. The circuit is basically the same as in the CTC 74, however, provisions have been made for width and high voltage adjustments.

A width coil, L404, is connected in series with commutating coil L401; a jumper across L404 may or may not be found due to production tolerance. Width can be increased by opening the jumper and if the raster sweep is too wide, L404 can be shorted.

The high-voltage adjustment switch, S401 in the primary circuit of T401, is not to be adjusted. This switch is usually sealed as shown in the diagram.

High-Voltage Protection Circuit

The circuit as shown in Fig. 3-32 looks like the horizontal "hold down" circuit in other XL-100 chassis. However, it should be noted that the circuit as used in the CTC 74, and CTC 81 chassis is not used to meet (HEW) requirements.

The circuit comprised of Q501 (and associated components) is designed to disable the horizontal sync circuits in case an open develops in the horizontal trace ITR circuit. The oscillator is forced off frequency, in the direction that prevents damage to components in the deflection circuit.

Under normal operating conditions (and most "failure" conditions) the circuit of Q501 is not conducting; cutoff state is maintained due to the reverse bias of the base-emitter junction—set by the 6.8-volt zener diode in the emitter. Quiescent "positive" voltage, developed via CR504 from the sweep pulse (negative going), is not sufficient to

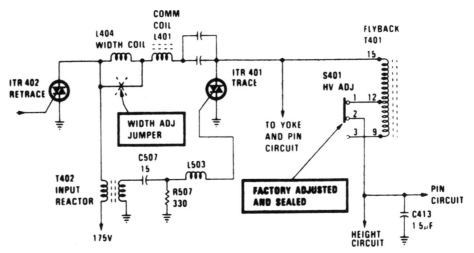

Fig. 3-31. Simplified horizontal deflection circuit.

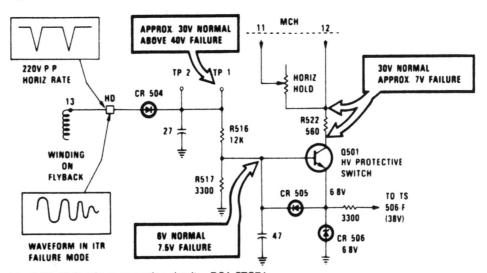

Fig. 3-32. High voltage protection circuit — RCA CTC74.

overcome the base-emitter junction. Under these conditions, the horizontal hold circuit (located in the MCH module) is unaffected, receiving its normal operating potential-the horizontal will remain in sync.

Note that the pulse available to the anode of CR504 (from terminal 13) is highly negative; in other chassis where this circuit acts as a hold-down-a "positive" pulse is applied at a comparable point, as a sampling of the high voltage being developed. Here, an unusual failure must occur in the deflection system, to cause a signal with "positive-going" excursions at the anode of CR504—sufficient to be rectified, fed to the resistive network of R516, R517 and the base of Q501—turning the transistor on. With this transistor biased into conduction, the dc potential applied to the horizontal hold control

89

(about 30 volts under normal conditions) is clamped to about 6.8 volts (CR506 in the emitter circuit). This voltage change drives the horizontal oscillator well off frequency, causing a loss of horizontal sync, protecting the deflection circuit until the fault is located and repaired.

Because its purpose is different, troubleshooting and testing the "protective" circuit in the CTC74 or CTC81 chassis requires a different procedure than used for the "hold-down" circuit in other XL-100 chassis. You may recall, the circuit of Q501 is normally at cutoff, having no effect on the horizontal oscillator. A negative going pulse is normally fed to the anode of CR 504 and, a pulse with high, positive going excursions— such as caused by an "open" failure in the trace ITR circuit—must be present at CR 504 to switch Q501 to an "on" mode.

Trace/Retrace Service Information

Possible damage to deflection components could occur if the trace ITR (or circuit) is defective. The protective stage is then defeated, and power is applied. Of course, a fault could develop within the Q501 stage, causing loss of horizontal sync; however, the original cause (or circuit) can be located by a proper sequence of servicing checks.

Assuming a loss of horizontal sync, the following service procedures should be used:

- Replace the MCH module with a good module, and readjust the customers hold control. If sync is restored at about the center of control range, a defective module is indicated.
- If, with adjustment of the customer's control, the oscillator can be adjusted close to sync, center the hold control and slightly adjust the horizontal range control on the MCH module to sync-in the picture.
- If conditions do not improve, measure the dc voltage at terminal 12 of the MCH module. If other than normal voltage is present at this point, replace the trace ITR, and recheck for correct operation, indicated by the return of sync, and/or a proper voltage at terminal 12.
- An abnormal voltage (6.8 volts) at terminal 12 of the MCH module confirms the "protective circuit" is "on", but does not indicate whether the origin is a defect, or caused by a component failure associated with the protective circuit (an emitter-to-collector short, for example). A dc voltage check at TP 1, TP 2, or the base element will indicate if the stage is being keyed on; viewing the waveform at point HD will yield the same result.
- A normal voltage at terminal 12 of the MCH module, indicates the protective circuit is not causing a loss of sync. The strong possibility now exists of problems in circuits associated with, or operating in conjunction with, the horizontal oscillator.

After any servicing operations, the protective circuit should be checked for proper operation. The value of some components, and the normal operating voltages in the

CTC 81 circuit are slightly different than in the CTC 74 circuit. The circuit as used in the CTC 74 appears in Fig. 3-32, while the CTC 81 circuit is shown in Fig. 3-33.

Operation in the CTC 74 can be checked out as follows: Set the line voltage to 120 volts ac and all controls for a normal picture. Momentarily short across R516 with a 12 k resistor. Under these conditions, the picture should lose horizontal sync, and the raster will become dimmer than normal. If these symptoms occur, the protective circuit is capable of working properly.

Operational checks in the CTC 81 follows the same basic procedure, however, R516 in the CTC 81 chassis is 18 k, requiring another value resistor to come up with a resistance of 6 k. A resistor substitution box can be used to shunt across R516. Loss of the horizontal sync should now occur under the above conditions. Voltages and waveforms during normal and abnormal operation should approximate those that are indicated in Figs. 3-29 to 3-31. The ringing, sine-shaped waveform (see Fig. 3-32) at point HD is representative of the signal with an "open" in the trace circuit.

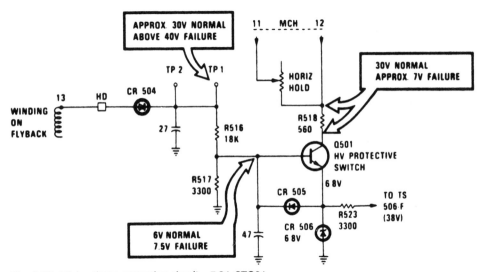

Fig. 3-33. High voltage protection circuit—RCA CTC81.

CHROMA SYSTEM (RCA SOLID-STATE IC'S)

The following IC circuits are used in the RCA CTC74 and CTC81 color TV chassis.

Chroma information is coupled from the MCK 001 i-f module to the MCC 001 chroma module (see Fig. 3-34) which amplifies and demodulates the chroma sidebands, providing color difference signals to the driver module MCD 001. The chroma module is supported by several external circuits such as the color and tint controls.

Some other inputs to the MCC 001 chroma module are horizontal rate keying pulses and blanking voltages for sets with digital address options.

Built within the chroma module is a regulated dc supply fed from the chassis 28 volt source and referenced to an 11.7 volt regulated supply derived in the luminance/sync module.

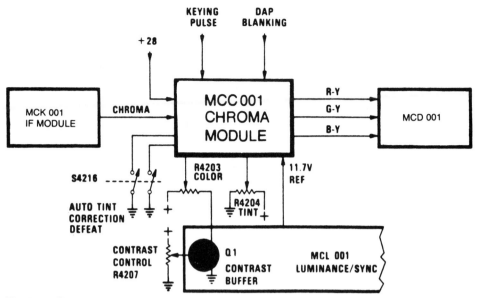

Fig. 3-34. Chroma module and related systems.

Chroma Module Circuits (MCC 001)

Looking at Fig. 3-35, note the chroma sideband information of about 600 mV enters the chroma processor integrated circuit, IC1. This circuit performs a number of functions including chroma amplification, automatic chroma control (ACC) and color subcarrier regeneration and control (AFPC).

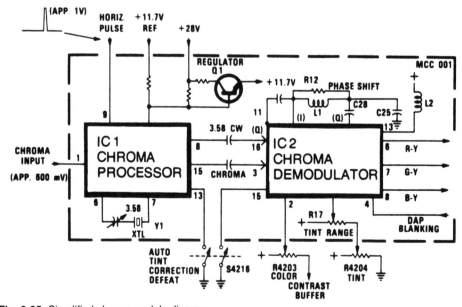

Fig. 3-35. Simplified chroma module diagram.

92

The exact frequency of the 3.58 MHz color subcarrier is set by crystal Y1 and capacitor C11. The color reference signal (burst) is separated by an internal gate activated by the 1 volt keying pulse that is fed to pin 9 of IC 1.

The processed chroma exits IC 1 via pin 15 and is capacitively coupled to pin 3 of the chroma demodulator, IC 2. The 3.58 MHz CW subcarrier information enters IC 2 on pin 16 at the proper phase to provide "I axis" chroma sideband demodulation. A phase shift network, consisting of R12, L1, C28 and C25 sets a 90 degree phase shift for the subcarrier to allow for "Q axis" demodulation. A customer color control, R4203 sets the gain of the IC 2 input chroma amplifier.

The subcarrier information may be phase shifted by an internal tint control circuit. The degree of phase shift is adjusted by the tint control pot, R4204. The demodulation process yields R-Y, G-Y, and B-Y difference signals at the proper respective amplitudes.

The regulator transistor, Q1, provides an 11.7 volt dc regulated supply for operation of the chroma module. Power is supplied from the chassis 28 volt dc source, but the regulator is referenced to the 11.7 volts in the MCL 001 module. This reference is used in order to maintain luminance to chroma tracking with contrast settings independent of supply voltage changes.

The Chroma Processor (IC 1)

A simplified block diagram of the chroma processor (IC 1) is shown in Fig. 3-36. Chroma sideband information entering pin 1 goes to a linear chroma amplifier which is gain controlled by an automatic chroma control voltage developed in the ACC detector. The amplified sideband information is then coupled to the AFPC and ACC detectors as well as the second chroma amplifiers.

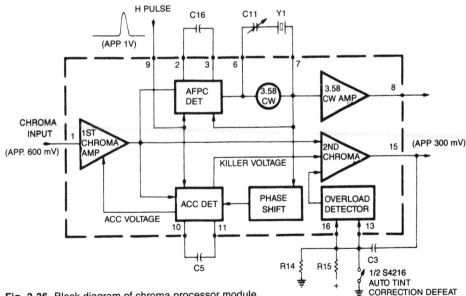

Fig. 3-36. Block diagram of chroma processor module.

93

The AFPC detector has the reference subcarrier (burst) gate and phase detector for phase locking the 3.58 MHz subcarrier oscillator. Capacitor C16 is an external filter for the AFPC circuitry while C11 and Y1 set the operating frequency for the oscillator.

The ACC detector is gated to allow burst input from the first chroma amplifier. Also applied is a phase shifted input from the 3.58 CW oscillator. This phase shift allows the ACC detector to become sensitive to amplitude variations in burst levels. These variations are filtered by capacitor C5 and fed back to the first chroma amplifier, providing gain control for this stage.

Loss of a burst signal causes the ACC detector to produce a voltage which cuts off the second chroma amplifier. Thus, the chroma circuitry is effectively "killed" when black and white picture information is received. The IC also has an overload detector which reduces the gain of the second chroma stage upon reception of an abnormally high saturated chroma signal.

The Chroma Demodulator (IC2)

In Fig. 3-37 we see the chroma demodulator, IC2. The chroma sidebands at about a 300 mV level are first applied to a gain controlled chroma amplifier. The gain of this amplifier is determined by adjustment of color control, R4203. Chroma blanking for digital address display (DAP) is also applied to this stage of the circuit. From the amplifier, the chroma sidebands are coupled to the "I" and "Q" demodulators.

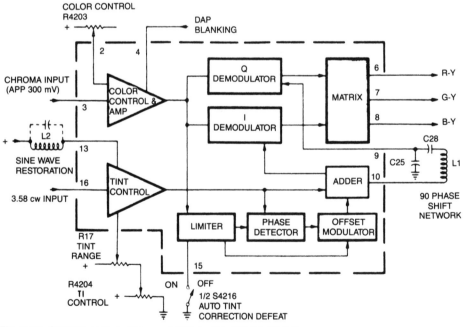

Fig. 3-37. Simplified block diagram of IC2, on the chroma module.

The 3.58 MHz CW subcarrier enters IC2 on pin 16 at a level of around 400 mV. The nominal phase of the subcarrier after being fed to the IC allows demodulation on the "I axis". However, the subcarrier phase can be altered by a dc controlled phase shifting (tint control) circuit. Control of this circuit is by the tint control, R4204, and a "tint range" centering control. Inductor L2 and its distributed capacity is broadly tuned to 3.58 MHz in order to restore a sinewave shape to the subcarrier waveform, which is distorted in the tint control circuit. The output of the tint control circuit is then fed to the "I" demodulator, via an adder stage, which has no effect on the subcarrier signal if the automatic tint correction feature is defeated.

A phase shifting network consisting of L1, C28 and C25 provides the 90 degree subcarrier phase shift for "Q" demodulation. The outputs of the "I" and "Q" demodulators are coupled to a matrix which provides the R-Y, B-Y, and G-Y difference signals at their proper amplitudes.

COMB FILTER CIRCUIT (RCA CTC 99/101 CHASSIS)

The purpose of comb filtering is to more perfectly separate the video signal and the chroma signal throughout the entire bandwidth of the video spectrum. Comb filtering greatly minimizes the cross-talk interference that would occur if chroma information were allowed to pass through the luminance channel. The comb filter operates on two basic principles.

- From one line to the next, the video information is basically the same. This means that as the picture is scanned vertically, there is very little change in the luminance information from one horizontal line to the next.
- Next, the chroma information on one line is 180 degrees from one line to the next, so the chroma information on one line is 180 degrees out of phase with the chroma information on the previous line.

Operation of the Comb Filter

The heart of the comb filter is U601 (see Fig. 3-38), the comb filter integrated circuit chip. This IC contains the charge-coupled devices and provides all of the functions of comb filtering: the delay, the amplifiers for the video and chroma processing channels, and the summing circuits.

The U601 IC is powered from several power supplies. These are: 16 volts B+ pin 22, 9.1 volts B+ pin 9, −5 volts B− pins 1, 5, 6, and 7. Video information is coupled into U601 at pin 11. The 10.7 MHz clock pulse, developed from the 3.58 MHz chroma subcarrier frequency tripler circuit, is inserted at pin 3. Pin 21 is the combed luminance output. Pin 13 is the vertical detail information output and pin 14 is the combed chroma information output.

To understand the operation of the comb filtering, the frequency spectrum of the video signal in an NTSC frequency interleaved color system must be understood. The video signals occur between 0 and 4.2 MHz. However, to provide color transmission, a

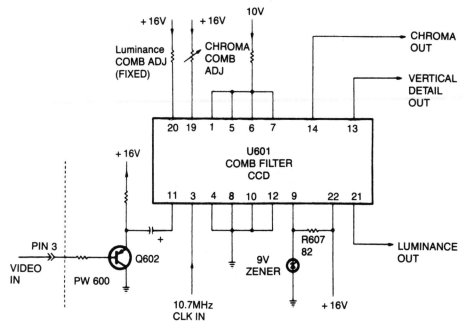

Fig. 3-38. Simplified comb filter diagram.

chroma subcarrier which is modulated by the color information is inserted at 3.58 MHz. The subcarrier, which is suppressed at the transmitter, creates sidebands which extend .5 MHz above the 1.5 MHz below 3.58 MHz.

To keep subcarrier sidebands from appearing in the luminance portion of the picture and generating cross-talk and other interference patterns, the video signal in a conventional receiver has a bandwidth of about 3 MHz. This minimizes the possibility of interference between low frequency chroma sidebands and high frequency luminance signals in the receiver's luminance channel. However, the limitations of video bandwidth causes a loss of high frequency (detail) picture information. Note video spectrum in Fig. 3-39.

Comb filtering allows the recapturing of video information between 3 and 4.2 MHz, thus restoring a large amount of picture detail.

The relationship between the transmitted video information and the scanning rate causes the video information to occur in "bursts" of energy at the horizontal rate multiples. Therefore, at every 15.734 kHz from 0 to 4 MHz a "burst" of energy exists containing the video information. The chroma information also occurs in "bursts" of energy at the horizontal rate intervals. However, because of the selection of the chroma subcarrier frequency (3.58 MHz), the energy bursts are offset from the video energy bursts by one-half the horizontal rate. Thus, the chroma information is "interleaved" with video information.

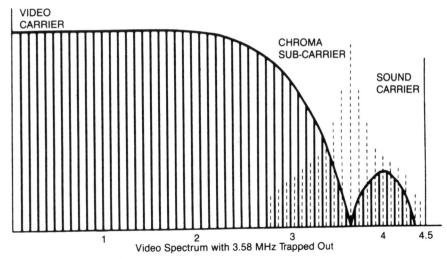

Fig. 3-39. Trapped 3.58 MHz video spectrum drawing.

Comb Filter Service Tips

Because the comb filter is just one of several circuits that process chroma and video information, the first step is to isolate the problem to the comb filter board or other circuitry, such as the i-f, chroma processor, video processor, and etc. In Fig. 3-40 you will find the key scope input waveforms and other checkpoints.

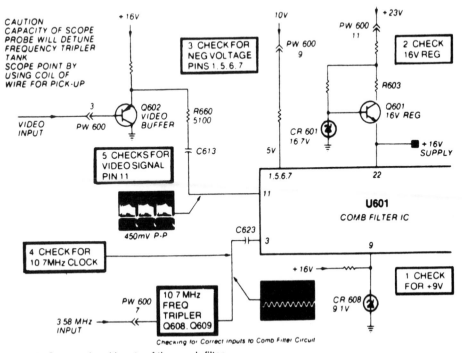

Fig. 3-40. Correct signal inputs of the comb filter.

- Check for proper B+. Pin 9 of the PW600 board should read −10 volts, and pin 11 should be +23 volts.
- Check for the presence of the video input signal at pin 3 of the PW600 board. If this signal is not present, the problem is in the i-f circuitry.
- Check for the presence of the 3.58 MHz CW at pin 7 of the PW600 board. This input is used to develop the delay line clocking pulses. Without this input, there will be no clocking pulses and the comb filter will deliver no output except noise. If the 3.58 MHz waveform is not at least 1 volt in amplitude, the comb filter will not be operational. If this voltage is incorrect, check the chroma circuitry for the defect.
- Check for chroma output signal at pin 1 of the PW600 board. If an output is present at this point, the problem is in the chroma circuitry.
- Check for video output at pin 15. If video output is present, the problem is in the video processing or video output circuitry.
- Check for the presence of the vertical detail output at pin 13. The lack of vertical detail output from the comb filter will appear as loss of picture resolution.

Loss of Video/Noise or Smearing in the Picture

Check the comb filter circuit U601 for +9 volts at pin 9 and for −5 volts at pins 1, 5, 6, and 7.

Check pin 22 of U601 for the +16 volts regulated dc power. If this voltage is not present, suspect regulator transistor Q601 or zener bias CR601 (16.7 volts) for the base of Q601.

Check at capacitor C623 for the 10.7 MHz clock signal. *Do not* connect the scope directly to the capacitor lead because the capacity of the scope probe on this circuit point will detune the frequency tripler. To check this signal, place a small loop of wire (or use scope pickup loop) on scope probe tip and coil it around C623. This will feed enough signal to the scope without detuning the tank circuit. Because of the capacity coupling, it is impossible to specify a peak-to-peak voltage. However, if the signal is present, this indicates the tripler is working and the comb filter circuit should now be operational. To tune the tank circuit, adjust coil L610 and L606 for maximum amplitude. After adjusting each coil, readjust the other to obtain maximum amplitude.

Check for presence of a video signal input to comb filter circuit at pin 11. If this video signal is not present, the problem may be in the video buffer Q602.

Now, with the scope, look for the combined video output signal at pin 21 of U601. If the output is not present, check dc voltage at pin 21. If it is not close to 4.6 volts, suspect a leaky filter capacitor or faulty U601. Also, check for the vertical detail output at pin 13 of the U601 chip. If this waveform is not present, there is a defect in IC U601.

Check the output of summing amplifier Q605. If this signal is not present, check the outputs at pin 21 and pin 13 of the U601. If the outputs are present at these two points and not present at the collector of Q605, then Q605 is defective. Likewise, if a waveform is present at Q605 and is not present at pin 15 of the module, the problem is in the buffer stage of the Q606 transistor.

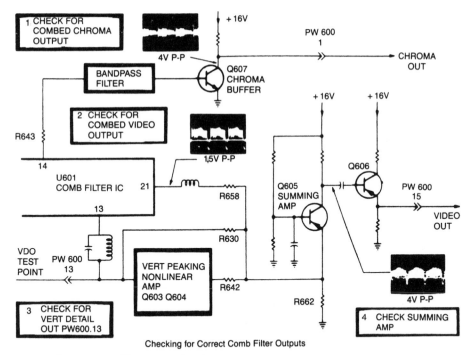

Checking for Correct Comb Filter Outputs

Fig. 3-41. Correct comb filter output signals.

Loss of Color/Good B & W Picture

Check the comb filter circuit U601 for +9 volts at pin 9 and for −5 volts at pins 1, 5, 6, and 7.

Check pin 22 of U601 for the +16 volt regulated supply voltage. If this voltage is not present, suspect regulator transistor Q601 or zener diode bias CR601 (16.7 volts) for the base of Q601.

Check at capacitor C623 for the 10.7 MHz clock signal. Use the scope pickup loop around capacitor C623 so as not to detune the tank circuit. If a signal is present, this will indicate that the tripler is operating and the comb filter circuit should be working. To tune the tank circuit, adjust coils L610 and L606 for maximum signal amplitude on the oscilloscope.

Check for the combed chroma output at the collector of the chroma buffer transistor Q607. If the waveform is not present at this point, check back to pin 14 of IC U601. If it is not present at this point, the IC is defective.

Notes for Comb Filter IC Replacement

If the comb filter IC is replaced, it is necessary to readjust the chroma comb adjust control. To do so, connect a scope to the vertical detail output point (pin 13) and adjust this control for minimum sync amplitude in the output waveform.

It is also necessary to check and possibly reset the frequency tripler tuned circuits and the i-f preset control R344.

I-F Preset Adjustments

- Connect channel 2 of a dual-trace scope to VDO test point which is pin 13 of PW600.
- Connect channel 1 of the scope to collector of bottom vertical output transistor Q504.
- Adjust the i-f preset control (R433) to provide 520 mV peak-to-peak amplitude of symmetrical VDO signal at the beginning of the retrace interval.

VIDEO CIRCUIT AMPLIFIERS (TUBE TYPE-AC COUPLED)

Many of these tube-type video amplifier circuits were used in small screen RCA color TV receivers. As we look at their operation, refer to the simplified schematic in Fig. 3-42 for the first and second video amplifier circuit stages.

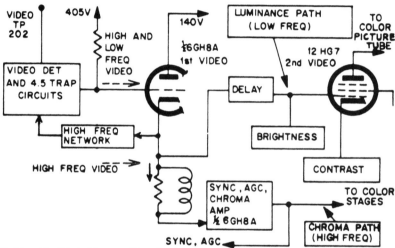

Fig. 3-42. Simplified first video amplifier circuit.

Video Circuit Operation

In these sets video detection is accomplished by a diode circuit and the detected signal is coupled to a two stage video circuit. The first video amplifier uses the triode section of a 6GH8A tube. This stage has a frequency-selective coupling circuit. Note that the grid and cathode are driven independently. The detector output video signal is fed in between the grid-cathode of the triode for the high frequencies. However, for the low frequency video signals, the triode acts as a cathode follower and provides a low impedance match for the delay line. The video signal then passes through the delay line to a second video stage that uses a 12HG7 high-gain video amplifier tube. The brightness signal is dc coupled from the plate of the second video amplifier to the cathodes of the picture tube via the green and blue drive control circuits. As in many other color sets a set-up or service switch is included for picture tracking adjustments. When in the set-up position, a fixed voltage is applied to the cathodes of the picture tube and the vertical sweep is collapsed. In this switch position the black and white tracking of the picture raster can now be easily performed.

Video Circuit Quick Checks

The first check when troubleshooting these video circuits is to measure all tube plate and screen voltages. For any voltages that are found to be improper, look for plate load resistors that have increased in value and leaky or shorted bypass capacitors.

For weak or complete loss of video use your scope to trace the video signal through all stages of the system in order to pin-point any circuit fault.

Components that are known to have failed in these circuits are the delay line, video peaking coils, and leaky or open coupling capacitors. Do not overlook faulty cathode bypass capacitors and open contrast controls. Any of these components may become defective intermittently.

VERTICAL BLANKING CIRCUIT (TUBE-TYPE)

This tube-type vertical retrace blanking circuit is found in many small screen RCA color TV receivers.

Operation of the Vertical Blanking Circuit

Vertical retrace blanking is accomplished in the plate circuit of the second video amplifier. Illustrated in Fig. 3-43 are the components associated with the vertical blanking operation. Let's now take a closer look at this vertical blanking operation. During active scan time CR702 is biased on (conducting); during vertical retrace time, a positive pulse from the vertical output transformer is fed through R756 to the diode. The

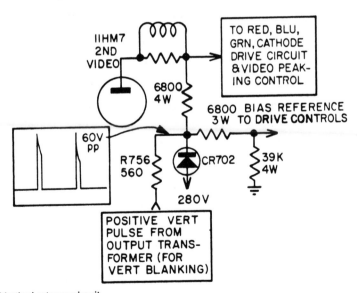

Fig. 3-43. Vertical retrace circuit.

60 volt positive pulse is sufficient to bias off the diode. During the blanking interval, the positive voltage pulse is added to the plate voltage of the 11HM7 tube and coupled to the cathode circuits of the picture tube, driving the cathodes more positive—cutting off the picture tube during retrace time.

Several important points should be remembered when signal tracing the video circuitry through the 11HM7 tube stage. The video signal at the grid of the tube is altered by the insertion of the horizontal blanking pulse. When the grid waveform is viewed at a horizontal rate on the scope, an additional negative pulse will be seen in the sync area of the waveform. Likewise, the plate circuit waveform (at a horizontal rate) will contain the added pulse during blanking time. The amplitude of the pulse at the plate will be determined by the setting of the brightness and contrast controls. When the plate waveform is viewed at a vertical rate, a positive excursion of the waveform will be noticed in the waveform area associated with the vertical blanking time. It should be noted that both the vertical and horizontal blanking areas will be altered from the output of the second video amplifier to the cathode of the color picture tube.

Points of Service

The most common faults in this blanking circuit would be loss of the vertical blanking pulses at the CR702 diode due to an open R756 resistor or defects in the vertical sweep stage. Also, an open or shorted CR702 diode will cause loss of vertical blanking and bright retrace lines will show up on the screen of the picture. Use the scope to look for low amplitude, distorted pulses or complete loss of these vertical blanking pulses in this stage.

COLOR DIFFERENCE AMPLIFIERS (TUBE-TYPE CIRCUIT)

These tube-type color difference amplifier circuits are used in many small screen RCA color TV receivers.

Circuit Operation

These color amplifier stages have several unique features as shown in Fig. 3-44. Note that the R-Y, G-Y, and B-Y amplifiers operate in the grounded cathode mode, with grid taken from the blanker circuit. Also, note that only capacitive coupling is used from the output of the difference amplifiers to the grids of the picture tube.

The dc reference level for the color grids is established using a clamp circuit. Each grid (red, green, and blue) has an associated clamp diode to provide the dc level for the grids of the picture tube.

Notice the symmetrical appearance of all three amplifiers up to and including the path to the picture tube control grids. The cathodes are grounded and grid-bias voltage is obtained from a divider network connected to the negative voltage available at the blanker grid circuit.

The plate output circuit of each amplifier has basically the same configuration. The connection from the blanker grid circuit to the common grid circuit of the difference

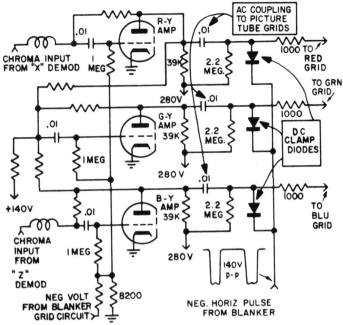

Fig. 3-44. Color difference amplifier circuits.

amplifiers sets the grid bias for the amplifiers. In other color difference amplifier circuits the blanker plate pulse is applied at the cathodes of the difference amplifiers, setting the operating point, restoring dc, and providing horizontal retrace blanking.

Key Points of Service

Check for correct B+ on plates of all color difference amplifier stages. If not correct look for plate resistors that have changed in value (opened or increased in resistance) and any leaky bypass capacitors or shorted spark-gaps that may be found in some circuits.

Shorted clamp diode will cause picture to be all one color. An open clamp diode may cause picture brightness to vary.

Check and make sure you have the proper negative bias voltage from blanker stage applied to control grids of the difference amplifiers.

If one color is missing, check for a faulty .01 μF coupling capacitor or open choke coil that feeds chroma information from the "X" and "Z" demodulators to the color difference amplifiers.

Make sure, with the scope, that you have the proper negative horizontal pulse from the blanker stage being fed to the dc clamp diodes.

TV SOUND CIRCUIT (SOLID-STATE IC)

This TV audio circuit is used in RCA's CTC-85 color chassis and several other models. The complete sound system is contained within one chip.

Circuit Operation

The 4.5 MHz sound information developed by the combination sound detector/ AFT discriminator chip in the i-f module is fed to the sound module circuit as shown in Fig. 3-45A. This circuit demodulates the 4.5 MHz sound signal and amplifies the recovered audio to a power level that will drive the 32 ohm speaker. The customer volume control also interfaces with this sound circuit.

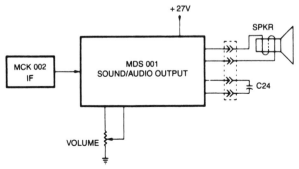

Fig. 3-45A. Block diagram of IC sound circuit.

As shown in Fig. 3-45B, 4.5 MHz sound and audio processing is performed in a single integrated circuit chip, U-1. For simplicity of explanation, the device is illustrated as two sections—"U-1A" and "U-1B". The 4.5 MHz sound information is coupled to a double tuned transformer (T1) via a coupling capacitor C1. It is then applied to section U31A which amplifies and demodulates the information. The audio information is then coupled to the audio preamp and power stages in section "U-1B" through R1 and C8. The customer volume control applies variable dc bias to the gain controlled stages in the preamp to allow adjustment of the audio level. Capacitor C24 (part of the speaker connector) provides the desired tone control characteristics. The power amplifier of U-1B furnishes a maximum power of approximately 1.5 watts (5% distortion) audio which is coupled to a 32 ohm speaker via capacitor C13.

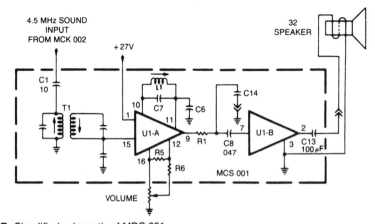

Fig. 3-45B. Simplified schematic of MDS 001.

Points of Service

For complete loss of sound check for +27 volts dc to the module. If this voltage is correct then check the speaker voice coil, as it may be open. The next check point is to see if you have correct 4.5 MHz sound signal input into capacitor C10.

For low or distorted sound suspect the U1 chip, the L1 coil or the T1 transformer. Also, check alignment of coil L1 and the T1 transformer.

SYNC SEPARATOR CIRCUITRY (SOLID-STATE)

This sync separator stage is used in the RCA all solid-state CTC-40 color TV chassis.

Circuit Operation

A simplified schematic of this sync separator circuit is shown in Fig. 3-46. Sync pulses appearing at the collector of the noise inverter stage are coupled to the sync separator amplifier. This stage provides a signal of the correct polarity to drive the sync separator circuit.

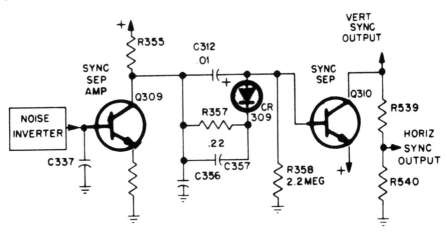

Fig. 3-46. Transistor sync separator circuit.

The sync separator is a PNP common emitter "switch" which is pulsed into conduction by the negative going sync pulses coupled to its base by capacitor C312. Resistor R358 provides a discharge path for this capacitor. The output of the sync separator consists of positive going sync pulses which are applied to the vertical and horizontal deflection sweep circuits.

The noise immunity of the sync separator system is optimized during horizontal sync time by the use of a relatively small coupling capacitor, C312 (.01 μF) between sync stages. However, at the frequency of the vertical sync pulses, the impedance of C312 is too large to hold the sync separator in saturation throughout the pulse interval. The initial part of the vertical sync is enough, however, to charge C312 sufficiently to forward bias diode CR309. This action effectively clamps capacitor C357 across C312,

105

providing a lower impedance path for the vertical sync pulses. Capacitors C357 and C356 provide filtering for high video and chroma frequency information.

Points of Service

Use the scope to trace the sync pulses throughout the sync separator circuit. Check for the correct vertical sync pulse output and the correct horizontal output pulse from this stage.

A defect in this stage can cause both the horizontal and vertical picture lock to be lost or just the vertical or horizontal lock to be lost.

Check for proper dc voltages to Q309 and Q310 transistor stages. Next, check transistors Q309, Q310 and diode CR309.

Look for value changes in resistors R539, R540 and leakage in capacitors C312 and C35.

AUTOMATIC GAIN CONTROL (AGC) CIRCUIT (SOLID-STATE)

This all solid-state (transistor) automatic gain control circuit is used in RCA's CTC-40 color TV chassis.

AGC Circuit Operation

The function of the AGC system is to maintain a relatively constant video detector output over a wide range of TV station input signals. Thus, changes in video signal amplitudes are translated into dc voltages which are used to control the gain of the RF amplifier and i-f stages.

Video information from the first video amplifier is fed to the gated AGC amplifier which produces a dc voltage proportional to the sync tip amplitude. The sync pulse is the only constant voltage level in the composite video signal that changes only when the signal input to the TV receiver changes. This dc AGC voltage is filtered and applied simultaneously to the RF AGC clamp circuitry and the i-f AGC inverter stage.

Gated AGC Amplifier

A simplified drawing of the gated AGC amplifier is found in Fig. 3-47. Video information with positive going sync pulses (level proportional to picture-carrier strength) is fed to the base of the gated AGC amplifier transistor. This transistor is made to conduct only during sync pulse time by positive going "keying" pulses coupled to the collector by capacitor C1. These keying pulses occur at a horizontal rate and thus key the transistor *on* simultaneously with the horizontal sync pulses contained in the video information fed into the base. The bias on the transistor is kept low to allow the base emitter junction to become forward biased only during the positive peaks of the sync pulses. This optimizes the noise immunity of the amplifier.

When the transistor conducts, electrons flow from the emitter to the collector, leaving a negative charge on capacitor C1. This negative voltage thus becomes the AGC

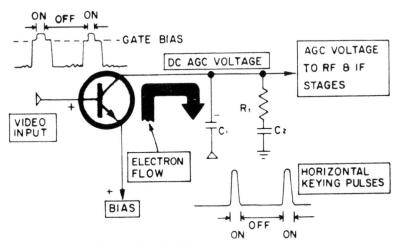

Fig. 3-47. Gated AGC amplifier circuit — simplified.

voltage. Its true value will now be proportional to the amount of amplifier conduction, which in turn is determined by the peak positive amplitude of the incoming sync pulses. The RC network consisting of R1 and C2 improves the overall stability of the AGC system.

A more complete schematic of the gated AGC amplifier is shown in Fig. 3-48. In order to produce a blank raster for picture tracking setup, a positive voltage is fed to the AGC amplifier by the action of the setup switch. When the setup switch is open this allows the positive 30 volt potential normally dropped across R330 to forward bias diode CR 308 and appear on the base of the AGC amplifier. This voltage will then saturate the transistor, producing a highly negative AGC voltage which cuts off the RF and video i-f stages. This action removes the video information from the screen, which in turn causes a blank raster. A diode, CR305, is placed in series with the collector of the gated AGC amplifier to prevent the developed negative AGC voltage from discharging back through the collector-base junction between keying pulses.

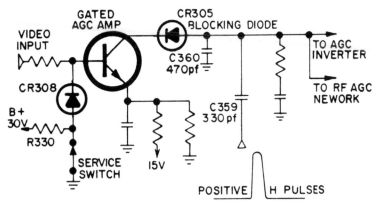

Fig. 3-48. Gated AGC amplifier circuit.

Key Checkpoints for Gated AGC Stage

Use a scope to check for proper video signal at base of the gated AGC amplifier stage.

With the scope check for the horizontal keying pulses that are fed to the collector of gated AGC stage via the C359 coupling capacitor. If the above two signals are not present or correct the gated AGC stage cannot operate properly.

Check for the +15 volt dc bias at the emitter of the gated AGC stage.

Suspect the following components if you have an AGC problem in this chassis: The AGC amplifier transistor, diode CR305, diode CR308 and coupling capacitor C359.

AGC CONTROL SYSTEM (SOLID-STATE TRANSISTOR)

This AGC circuit is used in the Sylvania E09 all solid-state color TV chassis.

AGC Circuit Operation

Referring to the circuit in Fig. 3-49 we see that this AGC system consists of video inverter stage Q414, AGC comparator stage Q416, buffer stage Q202, and tuner AGC inverter stage Q203.

Transistor Q414 is driven by noise free video from IC400. This stage inverts and reduces the amplifier of the video to provide proper drive for the AGC comparator, Q416. The AGC comparator stage collector is ungated, and powered by a floating dc voltage developed by T400 sweep transformer pulse winding at pins 1 and 2, diode SC460, and capacitor C452. This stage compares the negative most portion of the composite video signal (usually the tip of the sync) and the dc voltage on the emitter, determined by the AGC control setting, R900. The comparator, Q416, controls the amplitude of the current pulse to filter capacitor, C217, and likewise, the voltage across it. This voltage is buffered by emitter follower, Q202, and is used to control the gain of the i-f and tuner to maintain a constant tip of sync voltage at the base of Q416, the transistor comparator stage.

For signals from 10 microvolts to about 1.5 millivolts, the i-f AGC controls the gain, and the tuner AGC controls the RF gain. I-f gain decreases as the applied voltage from R217 increases, until diode SC210 is cut off, at this point i-f gain is at minimum. Now, the tuner becomes the gain control factor. The RF stage is an FET and requires a voltage range from 10 volts to about 1 volt for maximum to minimum gain reduction. Transistor Q203 provides the necessary inversion and level shifting for the FET tuner. Gain control above a few hertz or flutter is accomplished entirely by the i-f stages. The combination of R226, R227, R1250 and C1250 components eliminate most of the ac signal and removes the tuner from the ac control loop. Capacitor C216 and resistor R218, provide the path for the ac component when diode SC210 is cut off. In this AGC system, further ac gain reduction is available from the i-f beyond the minimum dc gain. This action is due to C215 and R215. These components are used to lower the ac loop gain at near maximum i-f gain.

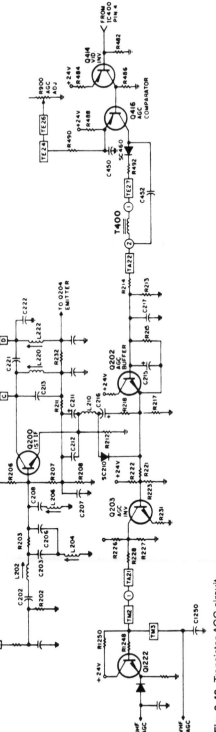

Fig. 3-49. Transistor AGC circuit.

Points of Service

Keep in mind that this is not a keyed AGC system. The pulses from the winding on T400 are only rectified by diode SC469 and the floating dc voltage applied to the collector of Q416, AGC comparator.

Check with the scope for a correct video signal at the base of Q414 inverter stage.

Check for faulty transistor and diodes. Then check for the proper dc voltages on these transistors.

Remember, that a fault in the i-f stages may cause a symptom that looks like an AGC trouble.

SYNC AND AGC CIRCUITS (TUBE-TYPE-ZENITH)

This Sync/AGC circuit was used in many models of Zenith tube-type color TV sets. This BA 11 tube circuit is probably one of the most interesting Sync/AGC systems in common use at this time.

Circuit Operation (Sync/AGC)

Referring to the 6 BA 11 circuit in Fig. 3-50, you will note a composite video signal, fed from the detector, has negative-going sync pulses and they are fed to pin 4. This signal usually is low in amplitude, and when noise free, has little effect on tube

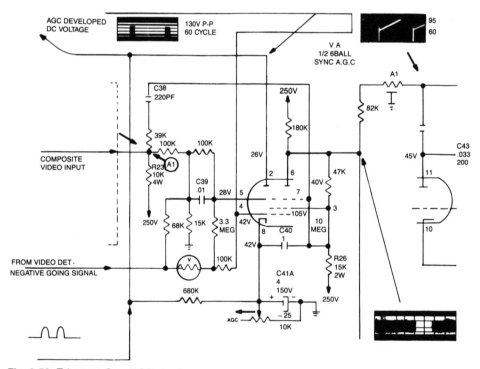

Fig. 3-50. Tube-type Sync/AGC circuit.

conduction. When noise pulses appear, a negative-going noise spike sends the tube deeper into cutoff for the duration of the noise spikes and prevents distortion of sync pulses at pin 6. The average conduction of the sync operating point is set so that only the upper 30 percent of the video signal brings the tube out of cutoff, permitting only sync pulses to appear in the output. Note that pins 3, 6, 7, and 8 are all common to the AGC section of the BA 11 tube. As the control in the cathode circuit is set for minimum resistance, grid bias between 4 and 8 goes more positive and the tube conducts more. Thus, the AGC voltage at pin 2 (plate) goes more negative and is applied to the video i-f grids, causing i-f cutoff.

When the AGC control is rotated clockwise, causing more cathode resistance, the bias is more negative and tube conduction is less. This reduces the negative AGC bias on the video i-f's, and will cause the i-f's to overload. During cutoff or overloading of the i-f's, video is blacked out and does not reach the sync-input grid at pin 7, which will affect operation of the sync-separator section.

Circuit Checks and Tips

The secret in troubleshooting this complicated circuit is to find out which section is defective, as one fault usually masks the other sections operations. The cathode (pin 8) and first grid (pin 4) are common to all sections of this stage. This first grid is part of the noise cancelling circuit. A good way to start checking is to short together pins 4 and 8 with a clip lead. If the set now operates, the trouble is in the noise cancelling circuit. Now check the video signal "backup stream" with the scope to find the problem. Also check the dc bias voltages. If the set still refuses to work, leave the jumper in place and check the AGC section next.

The AGC section is comprised of plate pin 2 and the two other grids on the left side of the 6 BA 11. Instead of a constant B+ voltage, this plate receives a pulse (keyer) from the horizontal sweep transformer. For AGC voltage to be developed, video information with sync pulses must be found at the grid pin 5. This part of the tube conducts only when the keying pulse at the plate coincides with the horizontal sync tip pulses at the grid. Thus, the AGC voltage is developed on the basis of the horizontal sync pulses amplitude. Use your scope to see that the horizontal keyer and sync pulses are present and correct.

For sync troubles, check for correct dc voltage at plate pin 6 and also scope this plate and grid pins 3 and 7 for the correct sync pulses.

A dual-trace scope is very effective for testing this sync and AGC circuit system. There are several video, sync, and keying pulses fed to and from the plates and grids of these circuits, and all pulses must be of the correct phase and amplitude or this stage cannot function properly. By using a dual-trace scope, two pulses can be monitored simultaneously for correct timing, phase, amplitude and waveform.

While adjusting the AGC or noise-cancellation control, and at the same time viewing these pulses on the scope, lots of service information can be obtained. If the set has no picture or sound, and if AGC trouble is suspected, the first procedure would be to clamp the AGC line. If there is any video still present, then proceed with the scope checks. If the set has no picture (no video signal to CRT grids) it is almost a sure bet there aren't any sync or video pulses present at the Sync/AGC tube elements. Should sync or video pulses not be found at the plates or grids, you may wish to inject pulses from a TV analyst in order to obtain some sort of picture on the screen of set under test. If partial operation can be attained, the scope can now be used to further pinpoint the faulty circuit stage or component.

ZENITH SYSTEM 3 ELECTRONIC
POWER SENTRY AND HORIZONTAL SWEEP SYSTEM

All of these new Zenith circuit features are located on the 9-160 module that can be repaired or replaced. The 9-160-03 module is used in the 13, 17, and some 19 inch screen size sets. The 9-160-04 module is found in most 19 inch sets and all 25 inch screen size receivers. As we look at and see how to troubleshoot some of these very complex and unique circuits you may wish to refer to the complete diagram of the 9-160-03 module that is shown in Fig. 3-51. This self-regulating horizontal sweep and power system provides ac line and load regulation without using a power transformer and is thus called an Electronic Power Sentry.

A block diagram of this Electronic Power Sentry system is illustrated in Fig. 3-52. A line operated 150 volt dc power supply provides the energy to a horizontal pulse-locked chopper. The chopper regulates the voltage and drives the sweep and High Voltage transformer with a constant pulse. The HV transformer also supplies all the regulated voltages to the rest of the set with a scan rectification scheme.

The "key" to the new regulation system is the pulse width modulation (PWM). The line and load voltages are sensed by the PWM which controls the chopper rate by varying the "on time" of the chopper, and therefore, the supply voltage.

The main point to keep in mind when servicing this chassis is that the voltage for operating the entire set are derived from the sweep transformer.

Circuit Operation

As the block diagram indicates, the 150 volts dc B+ is regulated by the pulse width modulator and horizontal output transistor. As the B+ increases, the PWM decreases the "on time" of the horizontal driver, decreasing the stored energy and keeping the sweep voltages constant. As the B+ decreases, the PWM increases the driver "on time", increasing the stored energy and maintaining a constant voltage at the sweep stage outputs.

The Chopper Circuit

As we look at Fig. 3-53 let's see how the chopper circuit is used in the electronic power system. Figure 3-53 shows a pulse-width-modulated energy storage step-up converter/regulator. The ac line connected bridge rectifier and filter provide a voltage to

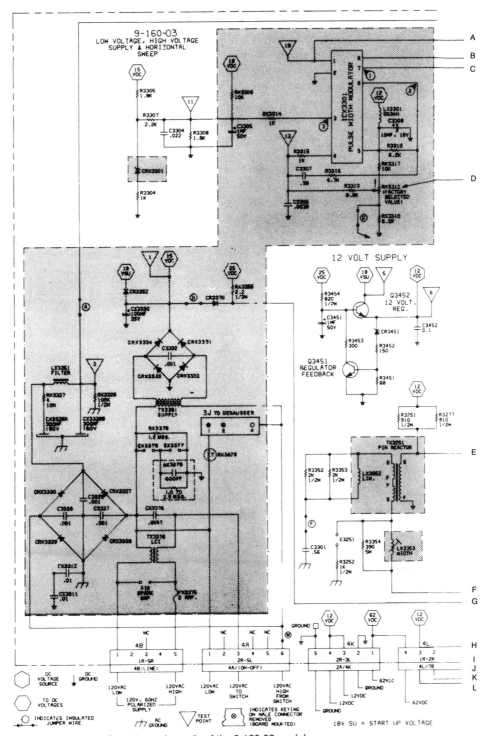

Fig. 3-51A, B & C. Complete schematic of the 9-160-03 module.

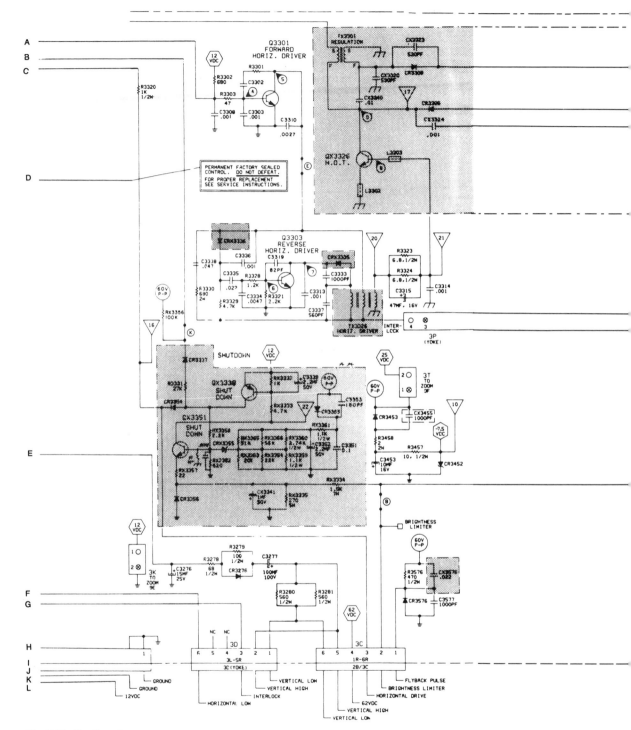

Fig. 3-51. Continued.

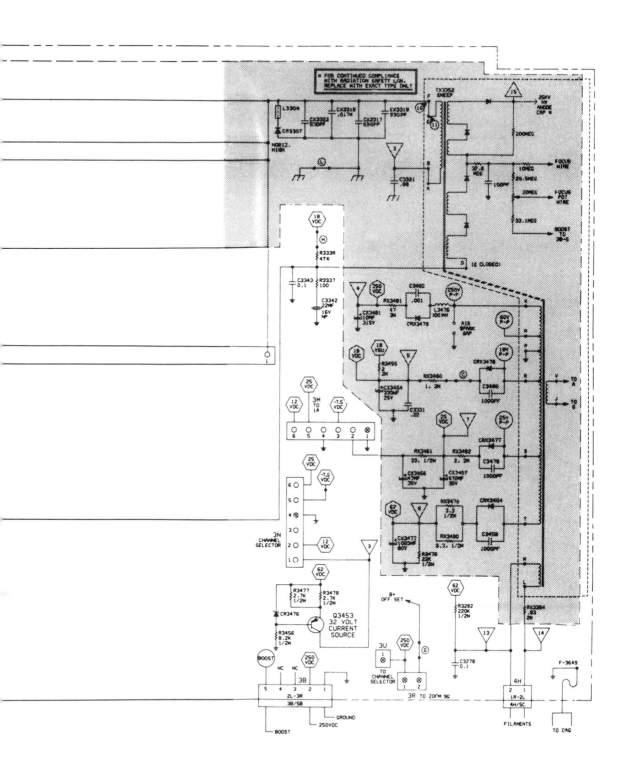

115

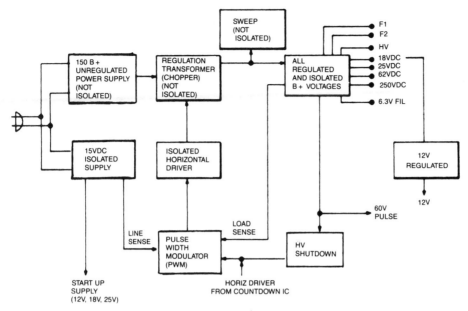

Fig. 3-52. Block diagram of the electronic power supply.

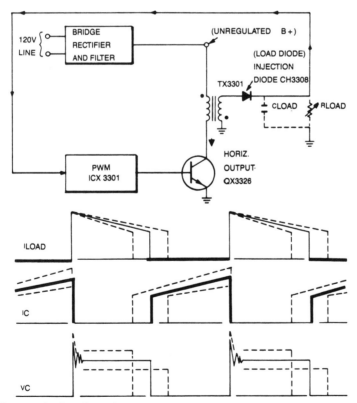

Fig. 3-53. Chopper circuit and waveforms.

the primary of the chopper transformer TX3301. The horizontal output transistor, QX3326, which is in series with the primary of TX3301, controls the stored energy by varying the "on time" inversely to changes in B+. If the B+ decreases, the "on time" must decrease, to maintain a constant current level. For a constant energy, current must be constant. The secondary winding of the transformer provides the step-up and phase inversion. During conduction of the transistor QX3326, the polarity across the secondary goes negative to positive. The negative voltage at the anode of the load diode CR3308 prevents its conduction.

After the QX3326 transistor turns off, the voltage on the primary reverses. Thus, the secondary voltage also reverses, forward biasing the load diode CR3308 and charging the load capacitance. The voltage at the load capacitance is sensed and fed back to the Pulse Width Modulator (PWM) ICX3301. The PWM varies the base drive, inversely proportional to the load voltage variations. This in turn provides the regulated load voltage. The peak "off" voltage of the transistor is also inversely proportional to the "on time."

Self-Regulated Sweep System

The circuits shown in Figs. 3-54 A & B are of the somewhat simplified regulator and horizontal output. The combination and further simplification of these two circuits evolved into the self-regulator, and is shown in Fig. 3-54C.

In the combined, evolved circuit, the transistor is removed from the sweep circuit, and will now operate as a chopper transistor. The load diode CR3308 is returned to the retrace capacitance. Since the drive power is supplied by the load diode, the B+ at the sweep decoupling capacitor, C3321, is not required. Since the chopper and sweep operate through a common transistor, there must be isolation during damper time. The isolation is provided by a diode CR3306 between the damper diode CR3307 and the transistor QX3326, the horizontal output. The diagram and waveform illustrations in Fig. 3-55 show the combination circuit in operation. After the completion of retrace time, the PWM is prepared to decide how soon the transistor will "turn on", to begin the energy storage process mode. For normal operation, the transistor turns on about 18 microseconds after retrace time, until the end of the trace. The "turn-on" time will vary from 5 to 30 microseconds after retrace. This time variation compensates for ac line to load variations as they occur.

The transistor QX3326 is "on" for the chopper operation during damper time. The diode CR3306 from the damper to the transistor, called the blocking diode, is reverse biased until the damper is off. When the damper cathode goes positive, the blocking diode is forward biased, clamping the yoke voltage via the transistor QX3326. The current through the transistor is a composite of the chopper current and positive yoke current. The transistor "turnoff" starts the retrace time.

Figure 3-56 shows the start of retrace. The transistor, QX3326, damper, and blocking diode are "off". The voltage at the collector of QX3326 rapidly goes positive due to the inductive kick of the regulator transformer. The polarity reversal for the

117

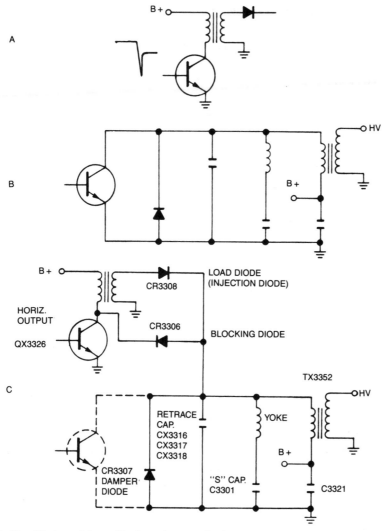

Fig. 3-54. Simplified regulator and horizontal output circuit.

primary of the regulating transformer is also seen on the secondary. The positive voltage at the anode of the load diode CR3308, referred to as injection diode, forward biases the diode into conduction.

The retrace is produced by the resonant circuitry in the deflection circuit. The injection diode CR3308 is forward biased because the anode voltage is increasing faster than the cathode voltage. The injection current aids in charging the retrace capacitance to a constant peak voltage level. This helps to add the lost power. Also, the injection circuit continues to supply energy during retrace time. Thus, the injection diode CR3308 continues to conduct until the pulse width modulator restarts the transistor with a requirement for more power.

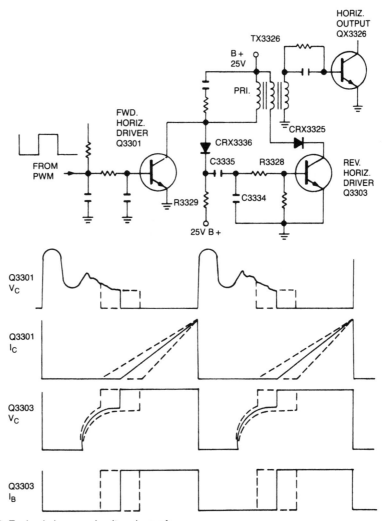

Fig. 3-55. Evolved chopper circuit and waveform.

The In-Phase Driver

The driver in the conventional sweep circuits is part of an out-of-phase system. For example, when the driver is on, the output transistor is off, and vice-versa. This system has an energy storage circuit; storing energy during conduction, and supplying output base drive during the off time of the QX3326 transistor.

In the self-regulation system, the driver is part of an in-phase system. When the forward driver Q3301 turns on, the output transistor also turns on. The driver circuit is powered by the 25 volt supply, which allows for a low resistance primary winding on the driver transformer TX3326. The closely coupled secondary winding is directly driven by a current ramp. The inductive impedance and low resistance provide for a very efficient driver circuit. The direct-drive produces a constant drive level independent of the duty

119

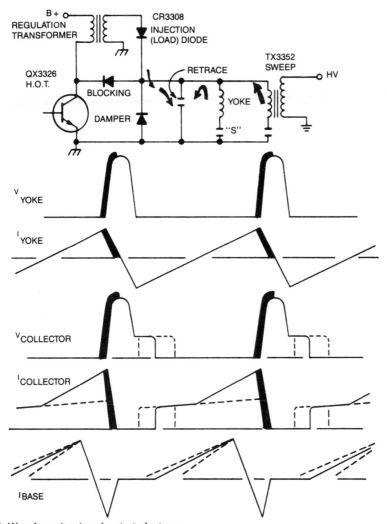

Fig. 3-56. Waveform drawings for start of retrace.

cycle. The conventional sweep circuit with the out-of-phase drive system requires a minimum "on time" to turn the output transistor "on" at low line voltages, and produces excessive drive current at high line voltages.

One disadvantage is the negative drive current which is unacceptable for a fast fall time in the output transistor. To maintain a fast fall time in the output transistor, a reverse driver transistor Q3303 is added to this circuit. The forward driver transistor Q3301 is driven by the pulse width modulator IC and the reverse driver transistor Q3303 is driven by the collector voltage of Q3301.

A diode CRX3336 and resistor R3329 form a clamp circuit which clips (or clamps) the peak of the collector voltage of Q3301. The waveform is decoupled by capacitor C3335. A capacitor C3334 and resistor R3328 form an integrating network which

120

shapes the signal that drives the base of Q3303. Transistor Q3303 turns "on" after Q3301 turns "off". The current through Q3303, flowing through the primary of TX3326, drives the base of the output transistor, QX3326, negative, providing a "clean" output pulse to switch QX3326 in about .5 microseconds.

The Pulse Width Modulator

Now, let's look at Fig. 3-57 which shows the schematic of the pulse width modulator and the internal circuit of the PWM IC. Pin 3 is connected to a reference voltage, while pin 4 is driven with a triangle waveform. When the triangle wave reaches a larger amplitude than the reference voltage, Q9 and Q10 will steal current from Q11 and Q12 from the current source Q13. The action of Q10's conducting will cause Q6 and Q7 (a positive feedback switch) to snap on. Transistor Q5 is an in-phase current mirror of the Q7 collector current. Therefore, Q5 will also turn on, feeding a large current into the base of Q3 which, in turn, will also overdrive Q1, the output device. The result will be a square wave output, starting and ending where the triangle waveform crosses the reference voltage applied to pin 3.

Let's dig in a little further now and see how this is used in the electronic regulator system.

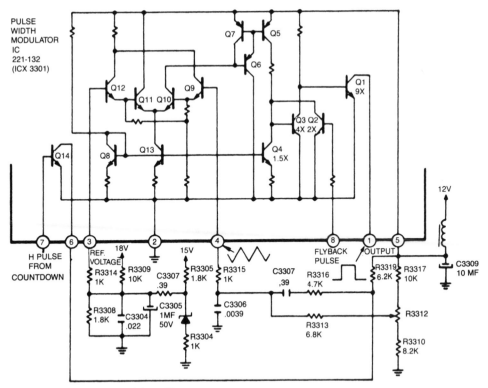

Fig. 3-57. Pulse width modulator IC.

121

Going back to Fig. 3-57, note that pin 7, of the PWM IC receives the "H" drive pulses from the "countdown IC" 221-103 located on the M2 module. When trouble-shooting this sweep system this is a good scope checkpoint to look at to see if these "H" pulses are present and correct. These pulses (note correct waveform in Fig. 3-59) are coupled to the base of Q14, an overdriven amplifier, which amplifies and clamps the "H" pulses between ground and B+, producing a 12 volt p-p square-wave. This square-wave exits the PWM at pin 6 and is then shaped by R3316, C3307, and C3306 to a triangle and coupled to pin 4, along with a dc reference voltage from R3317, R3312, and R3310 to the same point. On the other half of the differential comparator (pin 3) a sample of the voltage from the 15 volt, 60 Hz power supply, and a sample of the 18 volt sweep derived power supply are mixed. If either supply varies, the resultant square-wave output will vary in duty cycle, providing pulse width modulation. The actual trans-fer graphs for some typical voltage changes are shown in Figs. 3-58 and 3-60.

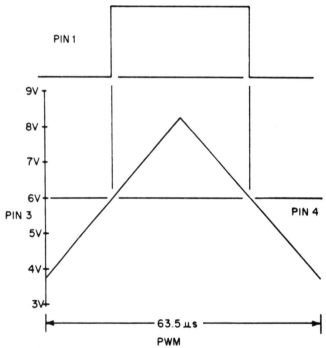

Fig. 3-58. Transfer graph.

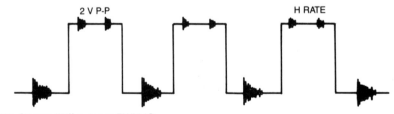

Fig. 3-59. Correct "H" pulse to PWM IC.

122

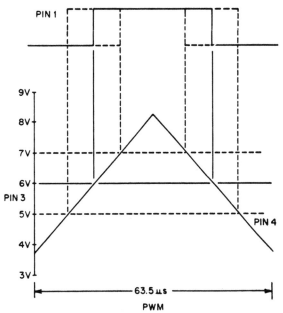

Fig. 3-60. Variations of transfer PWM characteristics.

To visually illustrate the action of the pulse width modulator, refer to the dual-trace scope pattern shown in Fig. 3-61. These waveforms were taken at the base of the forward driver transistor Q3301 which is indicated as scope checkpoint 4 on the full schematic shown in Fig. 3-51. The top trace occurred when the ac line voltage was at 130 Vac and you will note the pulse is narrow. Now the bottom trace widens out as the ac line voltage goes down to 75 Vac. Thus, a good example of the PWM in action.

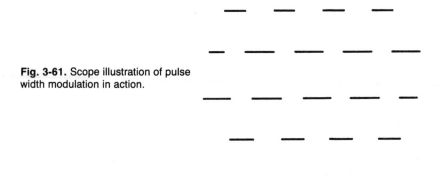

Fig. 3-61. Scope illustration of pulse width modulation in action.

Now, once again, return to Fig. 3-57. Look at pin 8 of the IC, which is marked "Flyback Pulse". This input can stop the square-wave from appearing at the output of the PWM at pin 1. It is used to prevent the PWM from turning on the driver and, hence, the horizontal output transistor from turning on, during flyback when the collector voltage is high. This input is also used to shut down the system if the high voltage becomes too high.

Ac Line Power and Kick Start Comments

This system has a 150 volt unregulated power supply and a 15 volt dc isolated power supply. All power for the receiver operation is derived from the unregulated ac line-connected 150 volt B+. This also includes the CRT filament. Don't try to measure any of the ac supplies from the sweep transformer. However, the dc voltages can be easily measured during any troubleshooting procedures.

The term "line-connected" B+, or "line-connected" ground, indicates no isolation from the ac line.

Keep in mind that the 15 volt supply is used only as a reference of the 60 Hz supply line voltage by the PWM at pin 3. The 15 volt supply is isolated from the ac line by transformer TX3351.

The power supply section is protected against lightning spikes or surges with an air spark gap, and a fuse for any short circuits that may develop.

High Voltage and Excessive Load ShutDown

A shutdown circuit is used for X-ray and HV sweep breakdown protection and goes into operation only if the high voltage increases by 4 to 6 kilovolts. Also, if an excessive load (current) exists or if the potential difference between the "hot" and "cold" grounds becomes excessive, its effect is to shut down the entire receiver by cutting off the horizontal stages.

This module contains a 12 volt regulator, a 32 volt current source, scan rectification circuits that produce 62 volts dc, 25 volts dc, 18 volts dc, 250 volts dc and voltage for the picture tube filaments.

Circuit Modification for Horizontal Output Transistor Failures

Remember, when a 9-160 module fails there can be various causes of horizontal output transistor failures other than a defective transistor.

One case, which is probably the most common, is intermittent or total loss of base drive. This may well be caused by poor contacts of the intermodule connector plugs or other component failures such as the 221-105 IC (loss of the 503 kHz oscillator), 221-103 IC (loss of the countdown process), or 221-132 (faulty PWM IC), or even components associated with these IC's.

Since this type of failure can be due to any of several other faults, a new circuit has been incorporated in the 9-160 module which protects the horizontal output transistor. However, if a horizontal output transistor has failed, check for an intermittent contact of the inter-module connector plugs.

In this new circuit, (see Fig. 3-62) a transistor (121-973) is placed across pin 8 of the pulse width modulator (PWM) IC 221-132 to ground. The transistor turns on when the 8.2 zener diode conducts. Thus, pin 8 is grounded through a 47 k resistor once start

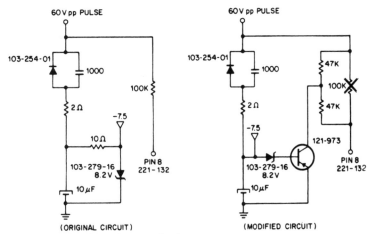

Fig. 3-62. Modification of horizontal output circuit.

up has occurred. If the horizontal drive signal should now be lost, the horizontal sweep circuit will squeg at a safe rate and will prevent overstress of the horizontal output transistor.

Another cause of horizontal output transistor failure is a loss of "off drive" which is the signal or current necessary for a fast cutoff time, incorrect or loss of the base current. The base current of the horizontal output transistor is due to the function of the reverse driver stage. When a horizontal output transistor is replaced, the base drive signal should be checked. The primary reason for loss of base current is a faulty reverse driver transistor or a reverse driver base clipper diode CR3336. It may be difficult to check the base current when the horizontal output transistor has been changed. So, it is important that at least the base voltage waveform be checked for correct shape and amplitude as the system may function properly for long periods of time before the horizontal output transistor fails again.

Another cause of output transistor failure may be caused by a breakdown of the heat transfer grease. In some cases when some black arc track marks are visible, the horizontal output transistor may be good and only the grease has broken down.

In one case where the output transistor shorted and blew the fuse, a customer removed the back of the seat and shorted across the fuse and turned the set back on to watch his special program. However, not for long, as a loud noise made him pull the plug, quick. This shorted transistor caused the C3315 capacitor to blowup and also burnt up the 6.8 ohm R3323 resistor.

Service Procedures and Tips (9-160 Module)

Use the following procedures for isolating the faulty module that will cause loss of (H) signal.

Turn the set on and check for presence of the 503 kHz (H) oscillator signal at pin 4 of connector 1B located on the 9-151 module. This will be a 1.5 p-p signal riding on a 2.5 volt dc level as viewed on the scope.

If this signal is incorrect or missing, replace the 9-151 module.

125

Now verify that the horizontal drive signal is present at pin 4 of connector 2B on the 9-152 module. This will be a 2 volt p-p signal riding on a 1 volt dc level.

If this signal is incorrect or missing, replace the 9-152 module. If all tests and modules have been OK up to this point then the 9-160 module is faulty.

Set Inoperative (Fuse Blown). When the fuse is blown on the 9-160 module, the cause could be the module or a short somewhere else in the receiver. A few checks with the ohmmeter should help locate the difficulty quite easily.

Make a resistance check on the Q3226 horizontal output transistor. From the collector (case) to the heat sink you should have a reading of 3k ohms or more. A very low or zero reading indicates a shorted device. Before repairing this module, make sure a failure of this output transistor was not caused by a defective 9-151 or 9-152 module.

Ohmmeter readings can vary due to a number of factors, such as the ohmmeter range used, the lead polarity, and etc. Always reverse the leads and use the lo ohms scale to make these resistance readings.

Excessive Picture Brightness (Retrace Lines). The cause of this symptom is usually the loss of the +250 volts on the 9-160 module (probable faulty diodes) or a defective 9-155 (RGB video module). In some instances a defective 9-155 module will also damage the 9-160 module. A faulty picture tube will also give nearly the same symptom.

To prevent damage to a new 9-160 module, make the following checks before module replacement.

Check for the presence of +250 dc at pin 3 of connector 3C on the 9-160 module. If the voltage is OK, check for +250 volts at pin 2 of connector 5B on the 9-155 module.

If the voltage is correct, then replace the 9-155 module. If no voltage is found, turn the set off and measure the resistance of R3481, a 47 ohm resistor. If the resistor is good, replace the 9-160 module. If you find the resistor open, replace both the 9-155 and 9-160 modules.

Deflection Yoke Tips and Checks. Should the horizontal windings on the deflection yoke become shorted, the screen will go black, the HV will drop to about half value and the 2 amp line fuse will blow. At first it may appear as a faulty 9-160 module. One quick check is to remove the 3P yoke plug and jump a clip lead from pin 4 of the 3P plug to pin 3 of the 3D yoke plug. Now turn on the set and the high voltage will go up and you will see a thin white vertical line on the screen if the yoke was defective.

With the yoke plugged in and a scope connected to pin 1 of the plug 3D (hot lead) and the ground lead of the scope to pin 6 of the 3D yoke plug, the waveforms taken at this point by the scope can be used as comparisons to give you an indication of the yoke's condition.

The bottom trace in Fig. 3-63 is for a good yoke, while the top waveform indicates an open yoke winding condition. While in Fig. 3-64 the bottom trace is for a good yoke and the top trace indicates a faulty yoke that has shorted windings and has loaded down and detuned the sweep output stage.

Another place where you can make a scope check in on the 250 Vdc supply pulse winding that is tap k on the TX3352 sweep transformer. A good pickup point on the

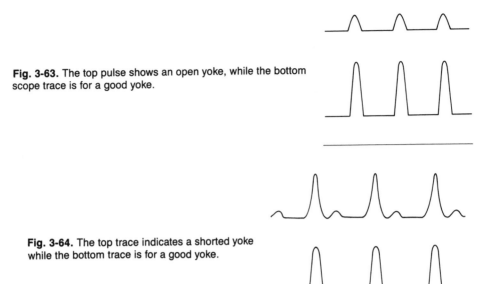

Fig. 3-63. The top pulse shows an open yoke, while the bottom scope trace is for a good yoke.

Fig. 3-64. The top trace indicates a shorted yoke while the bottom trace is for a good yoke.

board is at the L3476 choke coil. A good yoke will give a pulse like that shown in the bottom trace of Fig. 3-65. An open yoke will produce a pulse as shown in the top waveform. A shorted yoke will give a very low amplitude pulse at this test point.

High Voltage Shutdown Quick Check. To determine if the 9-160 module is in the shutdown mode, a quick and simple check can be made. Measure the dc voltage at pin 3 of connector 3C on this module. If the 9-160 module is in the shutdown mode a voltage of +8 to +9 volts will be measured at this pin. The normal dc voltage is +.9 to +1.2 volts dc.

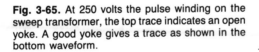

Fig. 3-65. At 250 volts the pulse winding on the sweep transformer, the top trace indicates an open yoke. A good yoke gives a trace as shown in the bottom waveform.

TV SOUND—SOLID-STATE IC (QUASAR TS-959 CHASSIS)

This IC sound system is used in the Quasar TS-958 and TS-959 color chassis. Most of the sound signal processing stages are within IC 201. This IC contains the sound i-f/Limiter, detector, and two audio amplifiers (preamplifier and driver) stages.

Circuit Analysis

Referring to the circuit in Fig. 3-66, we see that capacitor C201 couples the picture

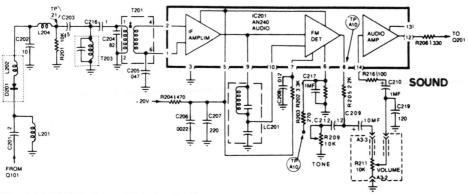

Fig. 3-66. IC i-f/Limiter and detector circuit.

and sound carriers from the third i-f stage to diode D201. These signals beat in the diode to produce the 4.5 MHz FM sound i-f signal which contains the transmitted audio modulations. This signal is fed through a selective tuned circuit to sound i-f transformer T201 which couples the sound i-f to pins one and two of the IC. The 4.5 MHz sound i-f signal is amplified and fed to the detector. The external quad coil between 9 and 10, shifts the phase of the sound i-f signal 90 degrees. This phase shifted signal is also applied to the detector stage in the IC. The detected audio signal is amplified and leaves the IC at pin 8. Capacitor C212 couples the signal to the tone control, and C209 couples the audio to the volume control. From the volume control, C210 couples the signal to pin 14 of the IC and to the audio driver section. The driver of the chip, pin 12, direct couples the signal to audio output transistor Q201. These preamp and audio output driver stages are shown in Fig. 3-67. Forward bias for Q201 is developed within the IC, and the amplified audio signal at its collector is transformer coupled to the speaker. A diac across the primary protects the output stage transistor against any current transients developed in the transformer.

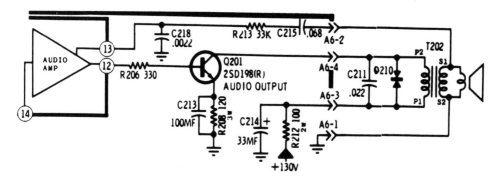

Fig. 3-67. Pre amp and driver audio stage.

Sound IC Checks and Tips

Voltage checks and/or signal injection techniques can be used when troubleshooting this sound IC and its associated circuits. First check all supply voltages to see if they are correct. Then check any other voltages in the associated circuits. Use caution to keep the meter probe tips from slipping and shorting any adjacent terminal pins. This can quickly ruin a good IC. If all voltages are correct, then inject a low-level signal at the input terminal pins, 1 and 2. Now, check the output of the chip with a scope for the audio signal. Another effective approach is to check the input and output of the IC with a scope for proper signals. These easy checks minimize the possibility of having to replace a good IC that could be damaged during removal.

VERTICAL SWEEP SYSTEM (SOLID-STATE TRANSISTOR) QUASAR

This transistor type vertical sweep system is used in Quasar TS-958 and TS-959 color TV chassis.

Circuit Analysis

The vertical sweep in this circuit is generated by altering the charge on capacitor (C457), very slowly during scan time and quickly reducing the charge during retrace time. The result is a sawtooth waveform that is presented to the base of driver transistor Q452 (Emitter Follower) which is passed on to the base of output transistor Q455.

Two NPN output transistors are stacked and share the B+ supply voltage. When in operation an average voltage (dc) develops across coupling capacitor C461.

The sawtooth voltage, applied at the base of Q455 controls its conduction, which inversely changes conduction of Q454. This type arrangement produces maximum current through the yoke at the top of the screen which decreases to zero at screen center then increases to a maximum in the opposite direction and moves the beam to bottom of screen. As the sawtooth cuts off Q455, the beam quickly retraces to the top and the process is repeated.

The complete vertical circuit Fig. 3-68, shows the positive feedback to the oscillator required to sustain oscillation. The same network provides negative feedback to the sawforming capacitor C457 to produce a more linear charge. Linearity correction is also enhanced by feedback from the emitter of output transistor Q455.

The amplitude of the sawtooth voltage developed by C457's charge is determined primarily by the height control (R456) adjustment.

The free running frequency of the oscillator is slightly lower than 60 Hz as determined by the vertical hold control. This allows the negative going sync pulse fed to the emitter of Q451 to lock the vertical oscillator to the TV stations transmitted sync pulses.

Diode D453 protects the base emitter junction of Q454 against excessive reverse bias voltages.

Circuit Troubleshooting Tips

Oscillation at the vertical rate is dependent on the feedback from the output. Anything that interrupts the path stops the proper charge/discharge of the sawforming capacitor. Voltage checks reveal most causes but a scope can be helpful in troubleshooting problems of vertical linearity and loss of picture sync.

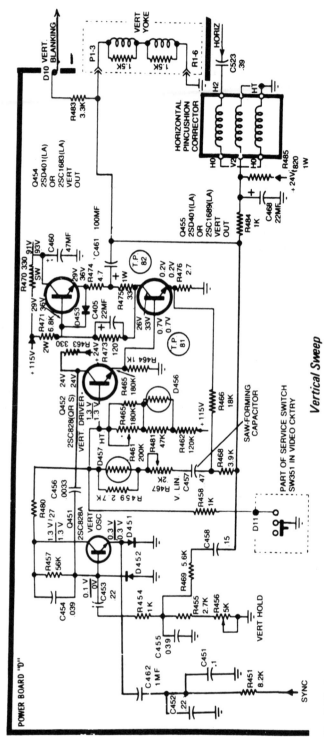

Fig. 3-68. Quasar transistor vertical sweep circuit.

Vertical Sweep

130

VERTICAL SWEEP CIRCUIT—SOLID-STATE (QUASAR TS-951 CHASSIS)

This solid-state vertical sweep circuit is found in the TS-951 and other Quasar color TV chassis.

During retrace time C408 couples the surge of positive voltage at the emitter of Q401 to the base of vertical driver transistor Q402. Positive voltage to this PNP transistor turns it off, and its collector voltage drops. Capacitor C409 couples the negative going voltage to the base of Q451, and drives vertical output transistor Q451 into cutoff. Its collector current drops to zero and the field in T402 collapses. This produces current in the secondary of T402 and onto the vertical windings of the deflection yoke. This vertical circuit is shown in Fig. 3-69.

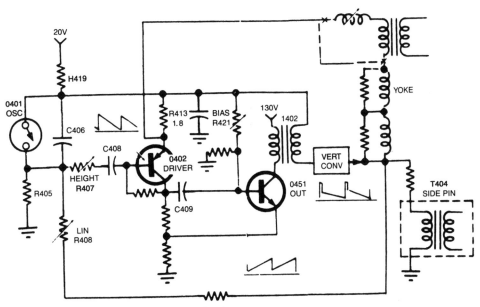

Fig. 3-69. Transistor type vertical sweep circuit.

At the end of retrace, oscillator transistor Q401 turns off and sawformer capacitor C406 starts to charge via R405. As this capacitor charges, voltage at the emitter of Q401 decreases, and C408 couples this negative going voltage to driver transistor Q402. The driver begins conduction and develops a positive voltage at its collector, which is coupled to the base of the output transistor to initiate conduction in this device. When Q451 begins conducting, current in the output transformer and deflection coils reduces slowly and the beams move downward to screen center. As the sawtooth voltage continues to increase conduction thru Q451, current in the output transformer secondary and vertical yoke coils reverses and deflects the beams linearly to bottom of the screen.

When the beams reach the bottom of the screen, the oscillator emitter voltage has dropped enough to let this transistor conduct and repeat the cycle.

It is essential that the free running frequency of the oscillator be slightly lower than 60 Hz, so vertical sync from the TV station will control it. Vertical hold control value determines this operating point. When a station signal is received C401 couples a negative going vertical sync pulse to the collector of Q401, and across the primary of T401. The secondary couples a positive going pulse to the base of Q401, which drives it into conduction. Height control R407 varies the amplitude of the sawtooth voltage to the base of the driver transistor, thus it controls vertical size.

Feedback from the vertical output transformer provides wave shaping to obtain good linearity. D402 and C412 protect the output transistor in the event the transformer secondary or vertical windings in the yoke open. The vertical bias circuit compensates for driver and output circuit tolerances, and R421 is adjusted to obtain best linearity in the top portion of the screen. This control should not need adjustment unless component replacement is required in this circuit for repairs.

ZENITH DIGITAL VERTICAL SYNC

The digital sync function eliminates seven external components and improves the locking characteristics of the circuit. Sync can be generated in two modes: Standard and Nonstandard.

Circuit Operation

Counter (G) detects the presence of Vertical Sync, by counting the clock pulses (1.79 MHz), during the sync period. Refer to Fig. 3-70 for the circuit diagram.

The first broad pulse, (6 μs) starts a window generated by flip-flop 2, which terminates after a period of 4 horizontal lines. If vertical integration does not occur, the counter proceeds to count out to the 833rd count and will terminate. During this period the counter is reset at each line and the broad pulses, (12 μs) are detected.

Counter (I) counts the number of 12 μs pulses during the window period. When two pulses are detected, vertical sync is generated by the (I) counter.

The leading edge of vertical sync occurs at the instant the second pulse is detected, and the trailing edge occurs at the end of the 833rd count.

If two pulses fail to be detected sync is not generated, and reset occurs.

VERTICAL SCAN

The saw waveform at pin 42 (IC 221-267) is developed by the charging and discharging of the .47 μf capacitor. Height control is at pin 3. Refer to Fig. 3-71 for the circuit.

Design accommodates large screen applications by adding diode CR 2150 at pin 2 and 3 of the vertical output IC 2100, and by adjusting the value of the yoke return resistor.

In the event a standard mode signal loses sync, the ramp will be generated by 525 counts at twice the horizontal frequency. This results in improved locking for weak signal conditions.

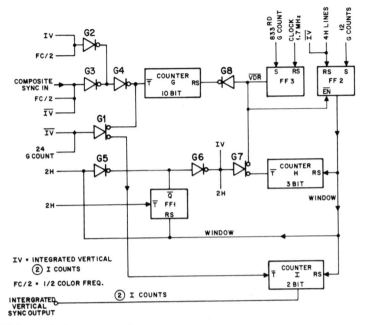

Fig. 3-70. Digital vertical sync separation.

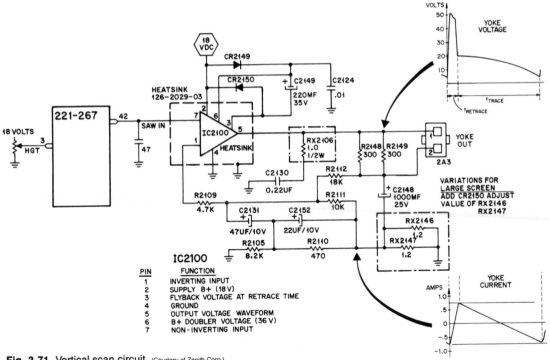

Fig. 3-71. Vertical scan circuit. (Courtesy of Zenith Corp.)

ZENITH HORIZONTAL SCAN

No external adjustment for horizontal frequency is required for this new IC 221-267. The free run frequency (15734 kHz) is automatically tuned eliminating a number of external components. Long term stability is enhanced under very tight parameters.

Circuit Operation

A master relaxation oscillator operates at 16 F, with continual compensation for B+ variations, thermal changes, and component aging. The block diagram is shown in Fig. 3-72.

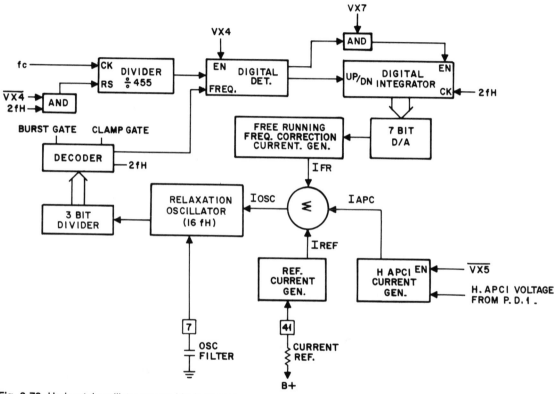

Fig. 3-72. Horizontal oscillator system block diagram.

Phase detector 1, is disabled for 6 lines during the vertical drive period. At this time, the oscillator frequency is compared to a reference frequency counted down from the 3.58 MHz color oscillator. If an error is indicated, the counters up-down state is altered and a correction is applied to the oscillator bias to correct the error.

The 16 F frequency of the master oscillator is divided down by eight and phase adjusted to center the picture at pin 9. A further 1 H division is applied to the second phase detector, which is compared to the flyback pulse. The resulting error signal from phase detector two modifies the horizontal signal. The corrected signal at pin 37, is the scan drive signal for horizontal sweep. Three advantages result from this system: Improved noise immunity, better picture centering, and burst and clamp pulses that are jitter free as a result of reference to the color oscillator.

BUCK TYPE REGULATOR

The circuit design in this Zenith set is a simple adaptation of the switch mode design that provides a nonisolated output with good regulation characteristics. Switch mode is at 35 to 45 kHz, assuring fast response to both load and line variations. Refer to circuit shown in Fig. 3-73.

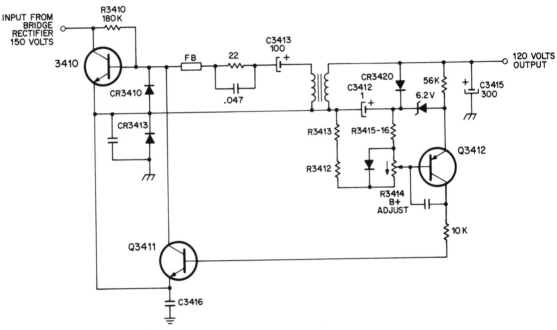

Fig. 3-73. Buck regulator circuit.

At turn on, Q3410 conducts and current flows in the transformer primary. As a result, secondary winding current brings Q3410 into saturation. At this point, current has increased to maximum and begins to level off. Transformer action ceases, cutting off Q3410.

With Q3410 cut off, the voltage across the transformer primary reverses its polarity, forcing current through CR3413 and the load, draining transformer energy.

The voltage in the transformer secondary now reverses in polarity, and the resulting current holds Q3410 in cut off until the primary current is at zero. This completes the switch cycle, which repeats at 35 kHz rate.

Regulation can be controlled by adjusting the amount of drive to Q3410. Drive sets the peak transistor/transformer current, which in turn determines the output load current.

135

When Q3410 is not conducting, the output voltage is across C3412 through CR3420. The resistance divider network, including the adjust pot, holds Q3412 off until the output voltage exceeds 120 V. At this point, Q3412 conducts to turn on Q3410 by shunting its drive. The overall result is a reduction of the load voltage.

The resulting regulator action provides tight control over the 120 volt load. The circuit references after each cycle, eliminating any tendency to overshoot control. An added plus for the regulator, is its behavior in the event of a short.

In a short mode, the regulator continues to operate, but at a low duty cycle. Q3410 conducts, but the transformer primary current is higher than normal, reaching a peak current much faster. Q3410 will be switched off, and the primary current clamped by the load short and diode CR3413. Transformer energy will be dissipated slowly and the secondary winding will not develop the reverse current required to reset C3413.

Repetition of the regulation cycle is dependent on capacitor reset through R3410.

The overall effect of the shorted load is a slowed down switching mode, with little power being dissipated. If the short remains, Q3410 will overheat and fail. If the load opens, the output voltage will tend to be close to the 150 volt input voltage, due to the circuit not being able to charge C3412. There will be no control of the drive bias at transistor Q3410.

ZENITH DIGITAL TV

The Zenith digital TV's consist of two primary versions. One is the standard digital set and the other is the Bose audio version.

Introduction

A digital TV was introduced by the Zenith Electronics Corp. in May of 1986. Let's now look at some general information about these digital TV receivers.

Over Thirty Years Ago. A digital TV receiver must process the same basic signal that was standardized by the FCC in December of 1953. The digital receiver must contain sync separators, chroma demodulators, etc. just as an analog receiver.

Baseband Video and Audio Signals. The baseband signals that modulate the video and audio carriers are recovered in their respective demodulators. A digital receiver will recover the same baseband signals. The circuitry of a digital receiver is basically the same between the antenna terminals and the video/audio demodulators as an analog receiver.

Output Stages. The four output stages of a digital receiver (video, audio, vertical and horizontal) operate in basically the same manner as those in an analog receiver.

A Major Conceptual Difference. The difference in the theory of operation between the digital receiver and an analog receiver is between the output of the demodulators and the input to the four output stages. The theory of operation involves changing the voltage level variations of an analog signal into an equivalent electronic number signal. For example, a maximum analog video signal variation from five to seven volts can be converted to an electronic number signal variation between zero and 265.

Zero could be equivalent to a level of 5 volts—256 to a level of seven volts, and the levels between five and seven volts to number values between zero and 256. This conversion process from levels to numbers is accomplished in a circuit called an *analog to digital converter* (ADC) in the digital receiver.

The familiar functions such as sync separators, chroma demodulators, etc. will be accomplished by manipulating electronic number signals. These number signals will be added, multiplied, subtracted, stored and delayed in digital circuits contained in IC's.

Reconversion. After manipulating electronic number signals to accomplish the functions of a color receiver, the numbers must be converted to the proper drive signals for the four output stages. For the video and audio signals this means applying the number signals to circuits called *digital to analog converters* (DAC). In this circuit every number value input signal will have a corresponding voltage level output. For example, a number signal input of 126 producing an output level of 6 volts.

To provide the familiar drive signals for the horizontal and vertical circuits, the electronic number signals are applied to digital circuits such as counters, registers, flip-flops, etc.

Different Circuitry. The components used to couple signals between circuits in the digital receiver are radically different from the coupling circuits in an analog receiver. In many cases in the digital receiver, there are no components involved in coupling a signal from one circuit to another. Just wire is used. This wiring can consist of four, seven or eight wires, with a common ground, that is used to couple a signal from one circuit (IC) to another. This multiple wiring for signal coupling is referred to as a bus and is used extensively in the digital receiver.

Computer Controlled Operation. The receiver functions in a digital receiver can be controlled by the execution of a computer program permanently stored in the ROM section of a dedicated microcomputer IC. Together with random access memory and electrically alterable read only memory IC's (RAM and EAROMS) this computer operation can store and/or change parameters that determine receiver operation. For example, traditional potentiometer settings or switch positions can be stored in memory circuits. These settings or positions can be changed (reprogrammed) by the viewer or service technician, to alter the operation of the receiver.

Advantages of Digital Receiver Design. There are a number of advantages to a digitally designed receiver, compared to an analog receiver. Even though it may not be economically feasible for a manufacturer to include these advantages in initial receivers, they can be included in future receivers. Some of these advantages are:

To top the list would be picture enhancement. This is accomplished by processing of the signals within the receiver that create the impression that government-established standards have been modified to produce better pictures. Increased numbers of scan lines and improved horizontal and vertical resolution of picture detail could be the results of enhancement techniques. Also, other features could include two channels simultaneously appearing on the screen (picture in a picture)—picture stop action (freeze frame)—storing a picture for later recall or print out (picture store)—reduction of the effects of multipath signal reception (ghost reduction).

ZENITH DIGITAL SIGNAL PROCESSOR/TUNER MODULE

The 9-535-02/03 is the "new" module in the digital receiver. It is new in the sense that many of the IC's on the module process video and audio signals after both signals have been digitized—changed from level variations to number signals equivalent to the level variations. It is not new in the sense that the signal processing functions such as sync separation, chroma/luminance separation, etc. that take place in an analog receiver must also be performed on this module. It is new in the sense that after the signal processing has been completed the number signals must be converted back to level changes and/or produce the four main color receiver outputs.

The block layout of the chip tuner and fourteen IC's is shown in Fig. 3-74. A brief description of the function/purpose of these chips will now be covered. The IC's used in the 9-535 modules is described in the Table 3-1 chart.

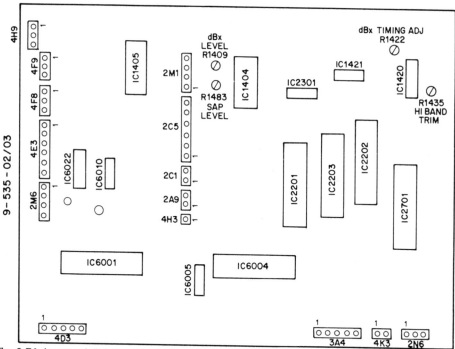

Fig. 3-74. Layout of digital main module.

The Central Control Unit (CCU)

The CCU is an eight bit dedicated microprocessor 40 pin chip. It contains a 6.5 k byte ROM section and a 120 byte RAM section. This IC has direct electrical connections to nine of the fourteen IC's. The program that is executed in the ROM section of the IC will indirectly affect the operation of four other IC's on the module. The IC is aptly named the central control unit.

Every time the receiver is connected to a source of ac power this IC is activated. The computer is reset and the program starts to run. The execution of the program

Table 3-1. Chart for the Main Module IC's and Function Type.

The main module 9-535 has 14 plug-in IC's. Referring to Figure C-1A, note the location of these IC's. All IC's except the ADC will come in a kit **905-87.**

IC6001	1)	CCU	Central Control Unit	221-479-03 or -06
IC2301	2)	MCU	Master Clock Unit	221-370
IC2201	3)	VCU	Video Codec Unit	221-373
IC2203	4)	NTSC	National Television Standard Commission	221-369
IC2202	5)	VPU	Video Processor Unit	221-371
IC2701	6)	DPU	Deflection Processor Unit	221-372
IC1405	7)	APU	Audio Processor Unit	221-376
IC1404	8)	ADC	Audio Analog to Digital Converter	221-377-01
IC6004	9)	TPPU	Teletext Processor Unit	221-365
IC6010	10)	EAROM	Nonvolatile memory chip	A-14343-05 or A14343-06
IC6022	11)		Tuner interface	221-250-02
IC6005	12)		Teletext Ram	221-364
IC1420	13)		Noise Reduction	221-303
IC1421	14)		Switching IC (audio)	221-188

NOTE: Where two part numbers are listed, the first is Non-Bose version and the second is the Bose version. The IC Kit 905-87 will include the 852-275 Digi-probe.

causes the CCU to set all the other IC's directly connected to it to be in their off state. The execution of the program waits for a POWER ON coded pulse signal to be applied to its pin 12. No other receiver operation can be performed until this signal is received.

When the POWER ON signal is received, the CCU will (1) distribute a reset signal to the IC's directly connected to it and initialize their operation and (2) after reset send data to and receive data from the IC's to cause the receiver to produce sound and picture.

There are two concepts that should be realized about the operation of the CCU once the receiver has been turned on. First is the fact that the CCU will respond to commands from the outside world—the viewer and/or service technician can input data from the keyboard or remote control unit to alter the operation of the receiver. *All of these command signals are coded pulse signals applied to pin 12 of the CCU.* These command signals cause channel changes, increase or decrease volume, color level, black level, picture level, etc. A service technician can input data to pin 12 that will (1) center the picture (2) set picture height and (3) set the free running frequency of the master oscillator in the TV receiver.

The second concept to realize about the operation of the CCU is the fact that it is constantly monitoring the operation of many of the circuits within the receiver. For example automatic B&W tracking, automatic brightness limiting, color oscillator drift, etc. are constantly being checked to maintain proper receiver operation. The data to maintain proper receiver operation is communicated between the CCU and other IC's by means of a three wire connection called the IM bus. The CCU can send or receive data from other IC's via the IM bus.

The Video Codec Unit (VCU)

The VCU is a forty pin IC and is the first IC to which video information is applied. It will accept two analog composite video input signals and an RGB video input. In the

Zenith application of this IC one composite video input (Pin 35) is the output of the receiver's video detector, and the other (Pin 37) is an external composite video from the "jack pack" at the rear of the receiver. The RGB inputs to the VCU are used for (1) teletext displays, (2) on screen displays and (3) RGB signals from computers.

The Audio Processor Unit (APU)

The APU processes the two 7.16 MHz pulse trains from the ADC into analog left and right audio output signals. This statement implies that some circuits in the APU must take digital data and convert it to analog signals and this is true. The circuits that convert digital signals into analog signals in the APU are called *pulse width modulators* (PWM). However, before the digital signals are applied to PWM's, several operations are performed on the digital signals.

Digital TV Servicing

When service is required on the digital television, some different and some familiar troubleshooting techniques will be used. Remember that the front end of the digital television is exactly the same as a normal TV set. There is a UHF/VHF combination tuner, 175-2264, and that works with a prescaler and microprocessor to develop i-f frequency that is detected to obtain composite video. This is normal television operation up to this point. The deflection circuits are conventional as with any other television set and can be serviced in the same manner as conventional televisions.

The main module has all plug-in IC's and will be serviced to the IC level. The only exception to this will be the 221-377-01 ADC chip. If this IC is defective the module will be replaced. The replacement module will have the ADC chip installed. Also mounted on the main module is the tuner and when the tuner is defective, the main module should be replaced.

The Zenith digital television consists of the following modules shown in Table 3-2.

Table 3-2. Chart Comparision of Non-Bose and Bose Main Module IC Types.

NON-BOSE	BOSE
1. 9-535-02 Main Module	1. 9-535-03 Main Module
2. 9-370-04 Switch Mode Power Supply (A-15125)	2. 9-370-04 SMPS (A-15125)
3. 9-387-02 i-f/Audio Module (A-13139-01)	3. 9-387-05 i-f/Audio Module (A-14161)
4. 9-367-07 Sweep Module	4. 9-367-07 Sweep
5. 9-369-03 Video Output Module	5. 9-369-03 Video Output
6. 9-410-02 IR Amp Module	6. 9-410-02 IR Amp
	7. 9-569 Bose Audio
	8. 9-553 Audio SMPS
	9. 9-648 Audio Input

4

Black and White TV Circuits

In this chapter we will be looking at circuits found in black and white TV sets, video monitors and microcomputer terminals. Most of these will cover circuit analysis and some troubleshooting tips.

VERTICAL SWEEP SYSTEM (TRANSISTOR) QUASAR TS-481 CHASSIS

The following circuits are used in Quasar B&W TS-481 and several other chassis:

Vertical Oscillator Circuit

The simplified circuit in Fig. 4-1 illustrates basic operation of the vertical oscillator. When power is applied, positive voltage from C304 drives TR32 into saturation. Any charge on sawformer capacitor C306 is shorted through diode D32 and TR32 to ground. At this time electrons flow from emitter to base and C304 charges to polarity shown in Fig. 4-1. Capacitor C304 charges very quickly and TR32 then cuts off.

The resulting rise of TR32 collector voltage, coupled to the base of transistor TR31, drives it into saturation. Note the complete vertical oscillator circuit in Fig. 4-2. TR31 then provides a discharge path for C304 through the hold control. As C304 slowly discharges (scan time) the base of TR32 becomes less negative (thus goes more positive), and forward biases TR32. When TR32 conducts, TR31 quickly cuts off, (retrace time) and the cycle repeats. The free running frequency of the oscillator is slightly lower than the 60 Hz vertical rate and is determined by the hold control. Just before C304 discharges sufficiently to allow TR32 to conduct, the negative sync pulse at TR31's base cuts it off and initiates conduction (retrace).

141

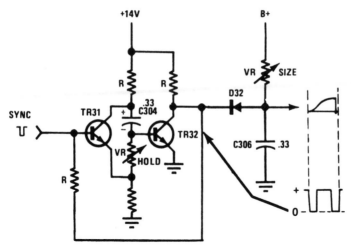

Fig. 4-1. Transistor type vertical oscillator circuit.

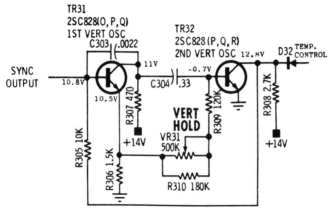

Fig. 4-2. Vertical oscillator and sawformer.

Sawformer Circuit

During scan time TR32 is cut off and capacitor C306 charges through the size control at an exponential rate. TR32 conducts during retrace time and discharges C306. The resulting sawtooth waveform is amplified by the driver and output stages.

Driver and Output Stages

When power is applied, output transistor TR34 conducts. This provides a path for yoke current from ground to B plus and the beam moves to the top of the screen. In normal operation a charge is accumulated on capacitor C311 and becomes the voltage source for output transistor TR35 and sawformer capacitor C306. Note driver and vertical output circuit in Fig. 4-3.

As sawformer C306 charges, driver transistor TR33 is brought into conduction. Its emitter is direct coupled to the base of output transistor TR35 and it also conducts. This

142

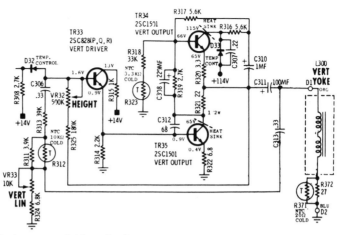

Fig. 4-3. Vertical output and driver circuit.

transistor's dropping collector voltage is coupled to the base of TR34 and reduces its conduction. Current through the yoke diminishes and the beam moves toward screen center. As the beam moves through screen center, TR34 cuts off, yoke current reverses, TR35 conducts more and the beam moves to the bottom of the screen. Yoke current for the lower portion of the scan is from the negative side of C311, through the yoke, R372 to ground, then back to R322, TR35 and R321 to the positive side of C311.

With the beam at the bottom of the screen, oscillator transistor TR32 again conducts shorting out the charge on C306. Driver TR33 and the output TR35 cuts off and output TR34 again conducts as the beam retraces to the top of the screen.

Linearity Correction Circuit

The normal charge-discharge of sawformer capacitor C306 through the resistance of the size control and R325 is at an exponential (nonlinear) rate. By injecting a sample of the output (which is the opposite phase) via C313 at the negative end of C306 the two opposite exponential curves, in effect, add together and cancel the nonlinearity. Control VR33 provides a means of varying the resistance from C306 to ground and thus alters the waveform produced.

Horizontal AFC Oscillator and Driver Circuits

In the AFC circuit (see Fig. 4-4) positive and negative sync pulses are applied to series connected diodes. A sawtooth (integrated horizontal pulse from the sweep transformer) is referenced to the horizontal sync. Any phase (time) difference between them develops a correction voltage that is fed into the base of the TR42 oscillator transistor. Horizontal oscillator and driver circuit operates as follows: Forward bias is applied from B+ to the base of TR42 horizontal oscillator via R406, R412, and R414. Collector current through the primary winding of L401 induces a positive voltage in the secondary, that is applied to the base through C410. This positive voltage quickly drives the

143

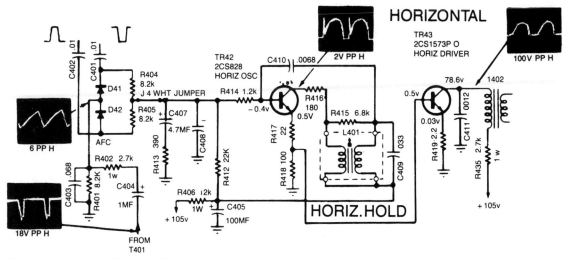

Fig. 4-4. Horizontal AFC and oscillator.

transistor to saturation and places a charge on C410. With no current change in the primary, the secondary voltage drops and turns off the transistor. The magnetic field then collapses. The transistor remains in cutoff until C410 discharges and forward bias is again applied. This above condition repeats itself at the horizontal rate as determined by L401, C409 and C410. The output is taken off the emitter of the oscillator transistor TR42 and amplified by the driver transistor TR43. This stage is transformer coupled to the horizontal output transistor TR44.

Horizontal Output Circuit

When the oscillator and driver turn on, the horizontal output cuts off and energy builds in L402. When the driver cuts off, this energy is released and the horizontal output transistor TR44 conducts. Collector current is from ground, through sweep transformer T401, emitter to collector of TR44 and R423 to the B+ supply. This creates a magnetic field in the sweep transformer and yoke, thus moving the beam to the right side of the screen. Note the horizontal sweep circuit in Fig. 4-5.

A negative going voltage applied to the base of TR44 (from L402) cuts it off. As the magnetic field in T401 and yoke collapses a large negative pulse appears at the emitter of TR44 and the beam retraces to the left side of the screen.

A negative 200 volt peak-to-peak pulse appearing at terminal 4 of the sweep transformer T401 is fed to the AFC circuit. This same point is used to supply positive 19 volts (after rectification) and is divided down to supply and power other circuits in the receiver.

Diode D49 rectifies the positive portion of the pulse (during scan time) and charges capacitor C491. If you have AFC problems check for a shorted diode D49 or C491 capacitor as they may cause this AFC pulse to be missing.

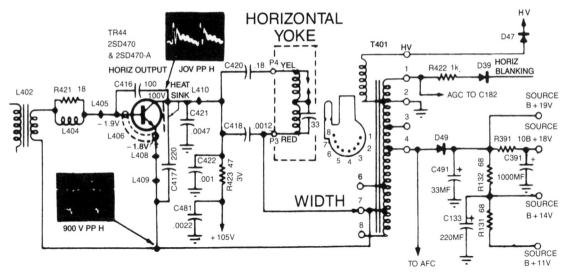

Fig. 4-5. Horizontal sweep output stage.

i-f AGC CIRCUITS: QUASAR B&W TV (SOLID-STATE)

The automatic gain control (AGC) system controls the gain of the first i-f amplifier to maintain a constant signal at the video detector. AGC is also applied to the RF amplifier under strong signal conditions to prevent signal overload, cross modulation effects, and etc. With no signal, or on very weak station signals, the i-f and RF amplifiers are biased for maximum gain. As signal level is increased the AGC system applies more forward bias to the i-f amplifier, thereby reducing its gain.

As signal level increases the i-f gain decreases until a predetermined level is reached. At a given level forward AGC voltage (more positive) is then applied to the RF amplifier to reduce its gain.

No Signal Condition

The AGC gate is forward biased by the positive voltage on its base and is normally in conduction. A positive going pulse developed during retrace time is coupled from pin 1 of the sweep transformer to the collector of TR18 gate transistor through C182, (.1 μF) and diode D18. Since the gate transistor is in conduction the upper end of C182 is clamped at a low dc level. As a result the high positive retrace pulse charges C182 as shown in Fig. 4-6. At the end of retrace the positive end of C182 is returned to ground via the sweep transformer pulse windings.

The negative voltage at the upper end of the capacitor reverse biases diode D18. It also opposes the positive voltage appearing at the junction of R186, R185 and CR181. This produces a greater drop across R186 and less forward bias appears at the base of the first i-f transistor TR11. This reduction in forward bias optimizes i-f stage operation for maximum gain.

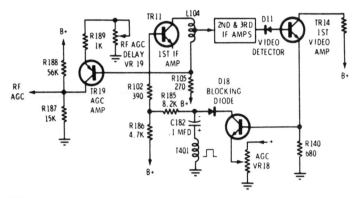

Fig. 4-6. i-f AGC transistor circuit.

The RF amplifier is also biased to produce maximum gain by voltage divider action of resistors R188 and R187.

When a station is tuned in, signals appear at the video detector diode D11. The diode detector recovers video information which is negative at D11 and this signal is fed to the first video amplifier transistor which appears on the emitter in the same phase. The emitter is dc coupled to the base of the AGC gate transistor TR18 (see Fig. 4-7) and this stage is normally in conduction. However, a negative going signal decreases the AGC gate transistor TR18's conduction (when the retrace pulse is present on the collector).

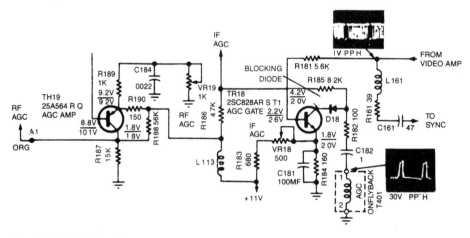

Fig. 4-7. RF AGC circuit.

With decreased conduction, C182's charge decreases, less negative voltage appears at the junction of C182 and R182 thereby allowing more positive voltage to be fed to TR11, first i-f amplifier's base. This increases conduction and the i-f gain is reduced.

146

RF AGC Operation

With strong signals, the first i-f stage TR11 is conducting harder, collector current increases and a larger voltage drop occurs across the 270 ohm resistor, R105. This drop is coupled into the base of the RF AGC delay transistor TR19. The large voltage drop across R105 appears as a negative going signal to the base of the transistor TR19. As a result it conducts (normally TR19 is cut off) and increases the AGC voltage divider that feeds AGC voltage to the tuner. With weak signals less voltage drop occurs across R105, TR19 transistor is cut off and the RF bias voltage is supplied from voltage divider action of R188 and R187. This action then causes maximum RF gain. From the above information it is now evident that the AGC system functions in a forward bias mode for the i-f and RF amplifiers.

SYNC SEPARATOR CIRCUIT (B&W TV SOLID-STATE)

The solid-state sync circuit shown in Fig. 4-8 consists of the sync separator stage and the sync amplifier stage.

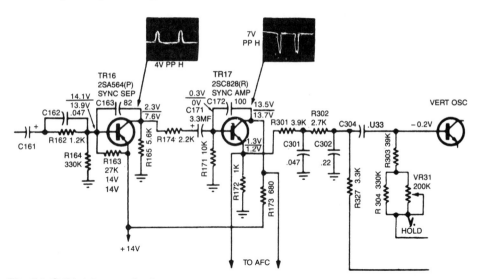

Fig. 4-8. Solid-state sync circuit.

Sync Separator Stage

The sync separator removes the negative going sync pulses from the video sync at the base of TR16. During sync time, emitter to base current charges C161 in the polarity shown. During scan time, the charge on C161 keeps TR16 cut off and effectively removes all other information. The negative sync tips turn transistor TR16 on and the resulting base to emitter current (during pulse time) restores the charge on capacitor C161. Amplified and inverted sync pulses will then appear at the collector of transistor TR16.

Sync Amplifier Stage

Positive going sync pulses from TR16 are coupled to the base of TR17, sync amplifier stage, via capacitor C171. The sync amplifier transistor TR17 operates split load with positive going sync from the emitter and negative going sync pulses from the collector.

The positive going sync pulses are coupled to the vertical sweep system through an intergrator network (R301, C301, R302, and C302) and then onto the horizontal AGC circuit through C402. Negative going pulses are then coupled to the horizontal AFC circuit through a coupling capacitor not shown in this circuit. When troubleshooting this circuit be on the lookout for leaky coupling capacitors, resistor value changes, and faulty transistors. Use the scope to trace pulses through these two stages.

VIDEO i-f CIRCUITS (B&W TV SOLID-STATE)

This i-f section consists of three discrete transistor common emitter amplifiers.

As we see in Fig. 4-9 the first i-f amplifier stage is AGC controlled and gain reduction occurs with increased forward bias. The i-f signal from the tuner is coupled via a coax cable to the i-f input (A3 and A4) terminals on the circuit board. The signal is coupled through a 39.75 MHz upper adjacent video trap, a 47.25 MHz lower adjacent sound trap and a bandpass shaping network to the first i-f amplifier transistor TR11.

Base bias for the first i-f is supplied from B+ through R186 (4.7 k) resistor and in conjunction with the AGC circuit, bias is varied in direct proportion to signal strength. Thermal stability is provided by emitter resistor R113 and it is bypassed to reduce degeneration.

The amplified signal at TR11 (first i-f) collector is coupled through L104 and C113 to the base of the second i-f transistor TR12. A 41.25 MHz trap at this point attenuates the sound carrier to establish a 10 to 1 ratio of video to sound carrier gain for proper intercarrier sound operation. The amplified signal at the collector is coupled through L106 and C115 to the base of transistor TR13, third i-f stage.

Base bias, on TR13, is established by R111 and R112. Capacitor C117 (4 pF) from the bottom of the collector coil to TR14 base couples an out of phase signal (180 degrees) that cancels out the signal fed back internally from collector to the base due to junction capacitance. The amplified output of this stage is coupled into the video detector through a capacitive network.

Capacitors C119, C120 and C121 resonate the coils in the third i-f collector and detector circuits. Capacitors C119 and C121 also provide coupling between the two stages.

AUDIO SYSTEM (SOLID-STATE IC)

In this system the sound i-f signal (4.5 MHz FM) is coupled from the first video amplifier's emitter to the audio IC via transformer L201 as shown in Fig. 4-10.

The first section of the IC provides both gain and limiting of the 4.5 MHz signal. The FM detector within the IC receives two 4.5 MHz signals from the limiter. One is

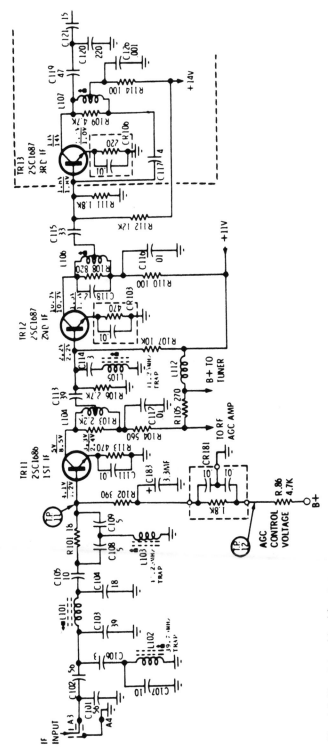

Fig. 4-9. Transistor video i-f system.

149

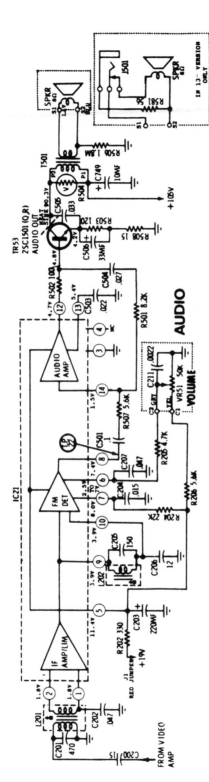

Fig. 4-10. Audio system — IC type.

applied directly to the detector, the other is shifted in phase by quadrature coil L202 (external) and returned to the detector. Frequency modulation of the i-f signal produces a phase difference (from 90 degrees) of the two signals reaching the detector and results in an output that corresponds to the original audio at the TV station. The recovered audio signal appears at terminal 8 of the IC and is coupled via C501 to the last IC section for low-level audio amplification. Control VR51 adjusts the dc bias of the FM detector output stages to vary the volume.

Output of the IC is fed to the audio output stage TR53, a class A amplifier. Emitter to base bias for output transistor TR53, is set by the DC level from the IC. Resistors R503 and R508 provide thermal stability and current limiting. Capacitor C506 prevents degeneration of low frequencies across R503. Transformer T501 matches transistor TR53 to the 8 ohm speaker. A VDR across the transformer protects transistor TR53 from high inductive kicks due to transients.

HORIZONTAL OSCILLATOR AND DRIVER (SOLID-STATE)

Forward bias is fed from B+ to the base of TR42 horizontal oscillator (see Fig. 4-11) through R406, R412 and R414. Collector current through the primary winding of L401 induces a positive voltage in the secondary, that is applied to the base through C410. This positive voltage quickly drives the transistor to saturation and places a charge on C410. With no current change in the primary, the secondary voltage drops and turns off the transistor. The transistor will remain in cutoff until C470 discharges and forward bias is again applied. The output is taken off the emitter of TR42 and amplified by the driver transistor TR43. The oscillator and driver stages conduct together.

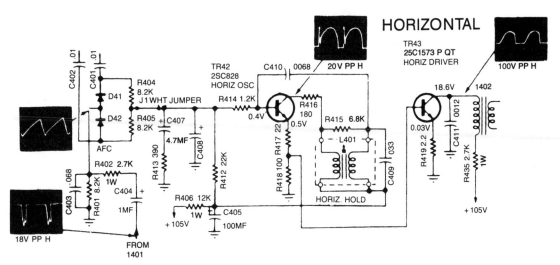

Fig. 4-11. Horizontal oscillator and drive circuit.

POWER SUPPLY CIRCUIT (SOLID-STATE REGULATED)

This regulated power supply circuit is found in many solid-state black and white video monitors or terminals. The complete power supply circuit is shown in Fig. 4-12.

Depending on monitor screen size and B+ voltage requirements along with the ac line voltage, the HI/LO T401 transformer secondary tap switch SW403, is used for nominal unregulated dc voltage from the full-wave bridge so that it may be maintained at about 23 volts for regulator efficiency. Set the SW403 HI/LO switch as follows:

The low position should be selected for line voltages of 90-110 (or 180-220) volts ac and the high setting for 105-132 (or 210-164) volts ac. The 115/230 volts ac switch, SW402, provides a series or parallel connection of two identical primary transformer windings for American or European ac line voltage sources.

A diode bridge rectifier unit is used to provide 23 volts of unregulated dc to the B+ regulator circuits. The main regulated B+ supply is generated or controlled by a series pass regulator transistor Q401 and R401 resistor driven by an error amplifier transistor Q403. An adjustable sample of the output B+ voltage is compared to the zener reference voltage.

POWER SUPPLY SERVICE TIPS (REGULATED)

The most common problems in this type of power supply are faulty rectifier diodes, zener diodes, regulator transistors and filter capacitors. A shorted bridge rectifier diode will open fuse F402 and an open diode will lower the B+ voltage and may produce a hum bar on the monitors screen. A short in the regulator transistors or filter capacitor C405B will blow the F403 fuse. Of course, any short up stream on the B+ line will also blow this fuse. If you find no regulation of B+ at all, suspect the pass transistor Q401, R401, Q402, Q403 and the zener diodes. An open C405A or B will cause reduced B+ voltage, hum bars and picture pulling. It may also cause picture rolling and weaving. An open R401, 25 ohm at 20 watt resistor, will cause complete loss of the 14 volt regulated B+ voltage.

HORIZONTAL OSCILLATOR (SOLID-STATE IC)

This circuit is used in black and white video monitors and display terminals for computer readouts.

Circuit Description

The horizontal oscillator and its frequency control is accomplished by IC801. Of course, the inner workings of the IC are very complex, but some circuit characteristics are determined by external components. As we dig into this circuit operation follow along with the diagram shown in Fig. 4-13.

For horizontal frequency control two input signals are compared and a resultant error voltage is fed to the oscillator to hold a frequency that is in sync with the input

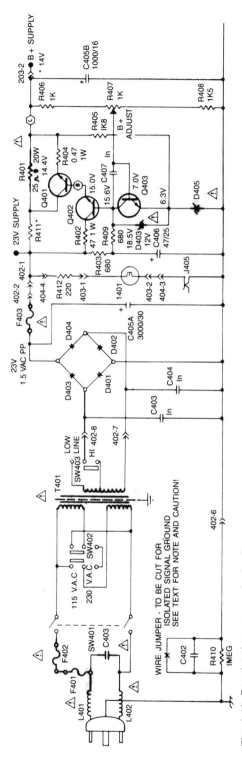

Fig. 4-12. Regulated power supply circuit.

153

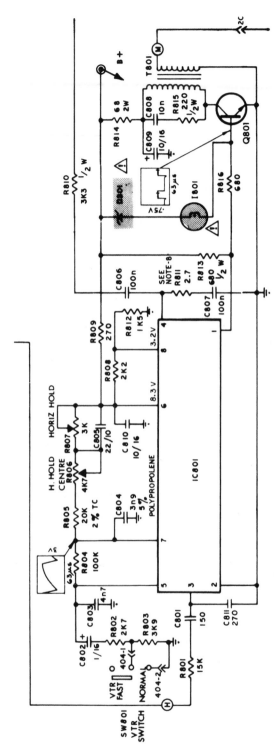

Fig. 4-13. Horizontal oscillator — IC type.

154

signal. The first signal is the negative going horizontal sync pulses which are fed to IC801 via R801 and C801.

A second 130 volt p-p feedback pulse from the horizontal sweep system is coupled via R810 to an integrating network. Resistor R912 provides a means of slightly shifting the reference sawtooth signal. This provides a method of shifting the raster for the best static relationship between the incoming sync and screen center.

The sync input pulse at pin 3 causes a sampling of the reference sawtooth voltage at pin 4 and the timing relationship will determine what dc voltage is developed when the sample occurs at the zero crossing of the sawtooth when no error voltage is developed and the input and oscillator are in sync. As timing varies an error signal is developed and fed to the external circuit at pin 5.

The oscillator time constant is determined by a thermally stable capacitive resistive network. The R807 pot is the adjustable horizontal hold control. R808 and R812 is an external divider network which controls the relative on/off time for the square-wave drive signal at pin 1.

The error voltage at pin 5 is coupled to the frequency determining network by R804 at pin 7 of the IC. The oscillator is corrected by this input.

The rate at which correction occurs is very important, especially during vertical retrace. If the horizontal pulses are missing during a portion of the vertical retrace, no error signal is developed and the oscillator will be free running. When the pulses return, and an error voltage is developed, the oscillator will be pulled back to the correct frequency.

Components C802, R802, R803 and C803 provide a time constant to maintain a control voltage during certain periods, such as the vertical retrace, when sync pulses are absent. Problems in this section of the circuit will appear as tearing at the top of the screen. Bending at the top of the screen with certain settings of the horizontal hold control occurs because the natural frequency of the oscillator is different from the sync and time is required to overcome the phase error. This time is dependent on the network time constants described already. For a faster response when using video tape recorders (VTR or VCR), SW801 (the VTR-VCR switch) removes the short across R803 increasing the circuits time constant.

Transformer T801 is required to provide a high current, low voltage drive to Q901, the horizontal output transistor. R814 and C809 decouple the transformer primary so that pulses are not transferred to the power supply, while providing the dc current for Q901. R815 and C808 were selected to reduce ringing in the primary of the driver transformer. Figure 4-14 shows the timing relationship of key voltage and current waveforms that occur during one horizontal scan and retrace cycle.

Troubleshooting Tips

If the oscillator is dead (no horizontal drive signal) check for proper dc voltages at pins of IC801 and the driver transistor Q801. Use the scope to check for a drive signal at the base of Q801. Also, look for a feedback pulse via R810 to pin 4 of the chip. Should

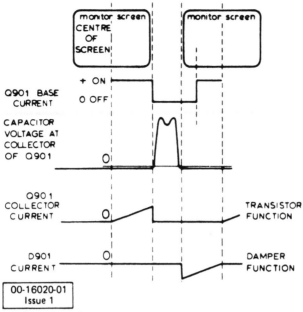

Fig. 4-14. Waveforms for horizontal scan system.

the picture not lockin (horizontally) use your scope to see if the correct sync pulse is at pin 3 of IC801. The most common problems found in these circuits is a faulty IC801, Q801 driver transistor and driver transformer T801.

POWER SUPPLY (B&W TV) AC/BATTERY TYPE

This circuit is found in the Quasar TS-483 B&W TV set that can be used on an ac power line or 12 volt battery. The set has a power transformer and is protected with a 1 amp fuse. The secondary winding provides ac to the full-wave bridge rectifiers. Capacitor C705 filters the 120 cycle ripple of the bridge output voltage. Regulator transistor TR73 regulates the B+ voltage at +11.5 volts. The circuit for this power supply is shown in Fig. 4-15.

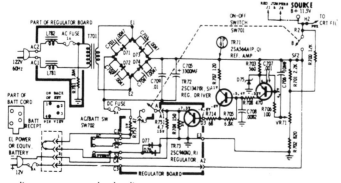

Fig. 4-15. Low voltage power supply circuit.

Regulator Circuit Description

Transistor TR73 functions as a variable resistance in parallel with R704. Its resistance changes directly with line variations and indirectly with the load.

The regulator circuit reacts so that an increase in output voltage reduces TR73's base bias and conduction, whereas a decrease in output voltage increases base bias and conduction. This results in a relatively constant output voltage with changes in line voltage and/or load variation.

Reference Amp and Driver

Zener diode D75 clamps the emitter of reference amplifier transistor TR71 at 4.8 volts. A sample of the output voltage from the divider network (R701, R702 and control VR71), appears at the base of TR71. If output voltage increases, its conduction decreases and the collector voltage drops (more negative). If the output voltage decreases, conduction increases and the collector voltage rises (less negative). The collector is connected through R708 to the base of driver transistor TR72, an emitter follower. Its emitter connects to the base circuit of regulator transistor TR73.

Charging Circuit

With the set turned off and the ac/Battery switch in the ac position the regulated supply charges the external battery when it is connected. The power supply operates as described above except that a jumper in the battery plug completes the ground leg.

A section of the on/off switch opens the circuit to the receiver section circuits. Resistor R700 connects from the reference amplifier base circuit to ground through the low resistance of the CRT filament to increase the output voltage for charging. The charging path is from ground through D76 (and D77) and the battery to the B+ output.

Battery Charging Operation

With the ac/Battery switch in the battery position, the negative lead of the battery connects through diodes D76 and D77 to ground. A 2 amp fuse to the positive battery lead protects the battery against overloads. You can use the troubleshooting procedures given in other sections of this book for regulated power supply systems.

SOLID-STATE i-f SYSTEM (B&W TV)

In this Zenith i-f circuit, during a "no signal" condition, the input AGC level is about +4.5 volts at test point "E" (Refer to Fig. 4-16) to the i-f and about +2.0 volts to the tuner, producing a very snowy raster. If no snow is observed on the screen (or at the second detector test point "C" with a scope), remove the AGC lead to the video i-f and insert a variable bias voltage, adjusted to +4.5 volts. If this opens up the i-f system and heavy snow is observed, there is an AGC problem: This indicates the tuner and tuner AGC are all functioning normally.

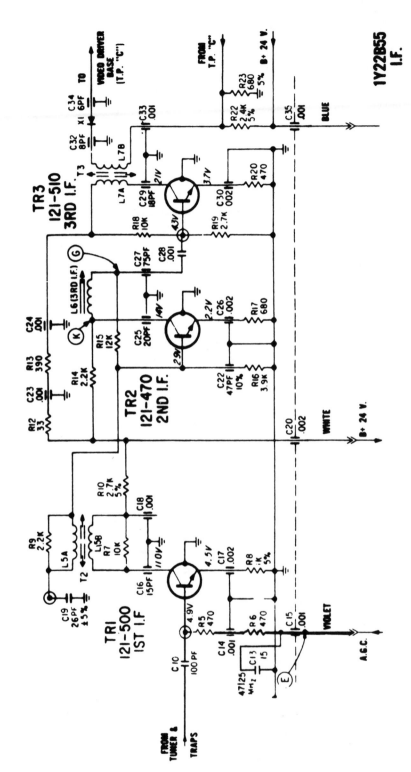

Fig. 4-16. Video i-f circuit diagram.

158

If some weak snow is observed on the raster, with the tuner disconnected, then the i-f amplifier stages are OK. If no increase in snow level is observed when the tuner is again connected, then there is tuner trouble: tuner AGC, transistor, or delay circuit. On the other hand, if snow on the raster comes up to normal when the tuner AGC lead is removed, then the trouble is in the tuner AGC delay since the tuner RF amplifier transistor (TR1) is automatically shifted to full gain with the AGC (delay) lead removed.

Thus, if the RF amplifier is excessively forward biased, (more positive at the base) it is still impossible to remove all snow, when the i-f amplifiers are good and operating at full gain.

Should no second detector signal be observed at test point "C" after making i-f bias tests, then check i-f transistor emitter voltages: stages are operating normally when (note Fig. 4-16) third i-f (TR3) is about +3.7 volts, second i-f (TR2) is +2.2 volts and the first i-f (TR1) will be on the order of +4.5 volts. In some chassis, without the AGC tuner delay transistor the first i-f emitter will be about +0.7 volt and the AGC output at test point "E" will be +1.5 volts under a "no signal" condition.

VIDEO AMPLIFIER CIRCUITS (ZENITH B&W TV)

The video driver (TR4) is an emitter-follower, with no gain, matching high impedance of the second detector to the low impedance of video output transistor (TR6), base. The collector of the video driver delivers composite video via the 0.22 μF (C50) and 1 k resistor to the base of sync limiter (TR10) for sync separator action. This video amplifier circuit is shown in the Fig. 4-17.

The video driver stage may overload and the second detector signal may measure 4 volts peak-to-peak but the output may be compressed or clipped at the 2 volt level. Thus, trouble is indicated in the output circuit. Check out the bias voltages and the video driver transistor itself.

It is possible to remove the Noise Gate Driver (TR5) action, by grounding the base as a troubleshooting check.

As a service check of the video amplifier system, use a 0.1 μF capacitor at 600 Vdc, to pick up the 6.3 Vac from the power transformer at pin 1 located on the CRT socket. This signal can now be used for signal injection to locate faults in the video stages as follows.

- Connect the capacitor through a clip lead to the collector of video output transistor (TR6), which should now show a little ac hum on the screen of the CRT.
- Next connect the capacitor to the base of TR6, and this will give you an indication of the transistor stage gain on the picture tube screen. The contrast control will have no effect during this test.
- There should be little change in ac amplitude when moving the capacitor to Test Point "J", at full contrast setting, with a properly operating Noise Gate Driver.
- Similar performance will be observed on the emitter of the Video Driver (TR4). Since the Video Driver is a no-gain device, similar amplitude should be observed on the driver base, test point "C", at the video detector.

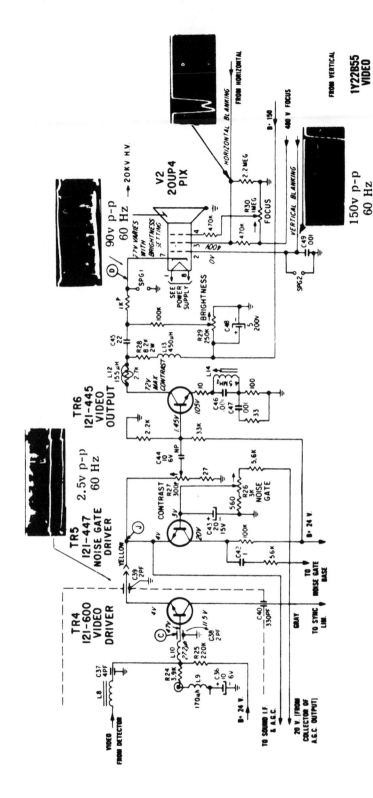

Fig. 4-17. Video amplifier circuit.

160

AGC CIRCUIT (SOLID-STATE ZENITH B&W TV)

This AGC system is "gated" with keying pulses fed from the horizontal circuit via 0.05 μF capacitor (C53). Test point "E" i-f AGC varies from +4 to +5 volts at minimum signal to +7 on strong signals. Tuner AGC varies from +2 volts to +4 volts (strong signal), with transistor tuner delay.

To adjust AGC control R33, (see Fig. 4-18), place the noise gate pot (R26) counterclockwise at minimum and adjust the AGC control just below the distortion point. Reset R26 just below the point at which picture position is affected.

R26 may also be adjusted with a scope as follows: Composite video sync tips should just be removed so that the stage will respond only to noise pulses exceeding the amplitude of the sync level.

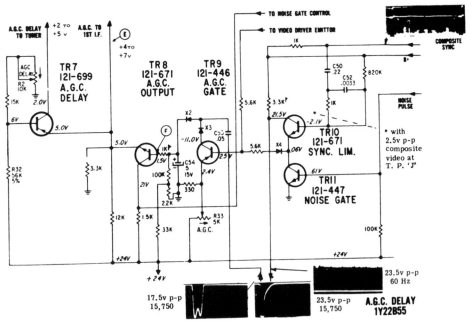

Fig. 4-18. AGC and noise limiter circuit.

It also is possible to set the AGC level for 2.5 volts peak-to-peak composite video information at test point "J". If AGC action is not correct, make the following checks:

- Short the noise gate (TR11) base to emitter, removing its possible effect on the sync limiter (TR10) and the AGC gate transistor (TR9).
- It is possible the noise gate pot (R26) is set too high, thus causing sync information to be clipped, and AGC action then occurs "off the video" instead of the sync tips which results in picture pull and roll.
- Watch for a shorted noise gate driver (TR5), emitter to collector. This will load down test point "J" and cause loss of video at the CRT.

- The AGC gate transistor (TR9) develops a bias proportional to sync tip amplitudes as a gated rectifier. This is fed to the AGC output transistor (TR8) at test point "F" whose emitter is used as follows:

  —Through test point "E" as i-f AGC to the base of the first i-f stage.
  —Directly to the emitter of the AGC delay transistor (TR7) and thus to base of the tuner RF amplifier stage.
  —A variable bias supply, positive going, may be connected to the emitter of AGC output transistor (TR8) and may be varied through the voltage range of +4 to +8 volts to simulate varying AGC levels, duplicating proper action for both the tuner and i-f AGC systems. This is a good technique to use for isolating AGC circuit troubles.

Diode X4 in the base of the AGC gate transistor (TR9) operates as follows:

- Under normal conditions the noise gate transistor (TR11) is biased to saturation, and the collector effectively shorted to ground.
- When the noise gate driver (TR5), in the video circuit, receives a noise pulse, it is amplified and fed to the base of the noise gate transistor (TR11), taking the stage out of saturation, thereby, lifting the collector off ground, which then affects both sync and AGC.

The AGC gate transistor (TR9) base bias is changed, causing it to momentarily cease operation for the duration of the noise pulse.

Diode X4 does not permit the sync limiter to go into conduction via the two series 5.6 k resistors and contrast control to ground, while the noise gate is out of saturation.

Note that during normal operation, the base bias of the AGC gate transistor (TR9) is determined by the voltage at test point "J", through the two series 5.6 k resistors via X4 to "ground" (the noise gate being in saturation or effectively shorted).

Diode X3 rectifies the gate pulses from the 0.05 μF capacitor (C53) to get the negative voltage required for the AGC gate transistor (TR9) collector.

Diode X2, between the AGC gate transistor and the AGC output transistor serves to isolate the gate pulses from the AGC output itself.

This 1Y21B55 Zenith B&W chassis uses the gating diodes instead of the AGC delay transistor stage. As we look at this circuit, refer to Fig. 4-19.

In the AGC output transistor (TR7) emitter circuit, under a weak or "no signal" condition; the voltage, at the base of the first i-f transistor (TR1) goes through test point "E" to the junction of diode X3 to the junction of the AGC output transistor (TR7) emitter (and 680 ohms to ground), maintaining the i-f AGC bias at about +0.7 volts for full gain.

As the AGC output transistor emitter voltage increases with increased signal level, X3 diode acts as a switch preventing test point "E" from exceeding the predetermined maximum voltage determined by the resistor divider action of R31 and R41.

162

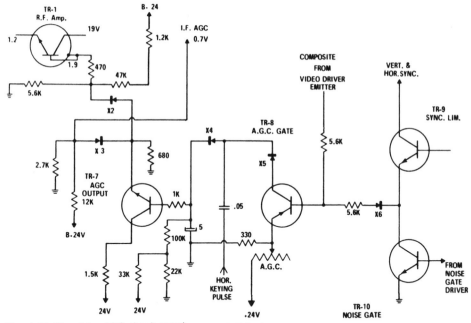

Fig. 4-19. Transistor AGC circuit-gated. (Courtesy of Zenith Corp.)

The tuner RF amplifier is biased at about +2.0 volts, determined by the resistor divider network of the 47 k and 5.6 k resistors. The way diode X2 is polarized, there can be no effect on the tuner RF amplifier base voltage until the AGC output transistor (TR7) emitter exceeds +2 volts. This is, however, about the voltage at which the first video i-f stage begins to approach a fixed bias on the resistor divider network R41 and R31.

At "no signal", or a very weak signal, the AGC output transistor (TR8) conducts little and the maximum i-f gain bias of about +4.5 to 5.0 volts is obtained from the 24 volts supply by divider resistors, with TR8 emitter at the junction of the two. VHF tuner bias is set at about +1.8 to 2.0 volts, at maximum gain, by the tuner resistor divider network. The tuner delay transistor (TR7) does not conduct at low signal levels.

As the input signal increases, the developed voltage at test point "F", base of the AGC output transistor, increases, it conducts more, and its emitter voltage (and the i-f AGC through test point "E") rises.

The tuner AGC delay transistor (TR7) emitter is biased at +2 volts, at the junction of R32 and the 15 k resistor between B+ 24 volts and the AGC delay control (R2) to ground. When the i-f AGC voltage rises, moving the emitter of the PNP tuner delay transistor (TR7) more positive with respect to the base, TR7 begins to conduct, and the tuner AGC voltage rises. At this signal input level, the i-f AGC voltage increase slows down and most of the additional gain reduction is done with an increasing AGC voltage to the tuner from the emitter of TR8, through the delay transistor TR7.

If AGC is not capable, alone, of providing sufficient reduction in gain under strong signal conditions. To provide sufficient reduction, AGC is applied to both the first i-f and the VHF tuner when the incoming signal reaches a certain level. By delaying the tuner AGC, maximum performance is assured on weak signals; the tuner delay entry point is determined by how soon the delay transistor TR7 begins to conduct. The AGC delay control sets the base voltage and thus the i-f AGC level at which the delay transistor, TR7, conducts.

The delay point is factory set, using R2, so that tuner AGC reads slightly more than +2 volts when input signals are between 1 and 2 millivolts. If R2 is adjusted too low (clockwise), tuner AGC action occurs early, and more snow may appear at low or medium signal levels. If, on the other hand, R2 is set too high (counterclockwise), cross modulation interference may appear under strong signal conditions. A good compromise may be obtained by setting R2, on a medium signal level, clockwise until a noticeable increase in background snow level occurs then just back off until the snow level decreases. The AGC level control setting of R33 is not affected much by adjustments of the delay control.

Sync Circuit Operation and Service Tips

The overall performance of the sync limiter (TR10) can be best checked out with a scope and a LC probe. Composite video will be viewed at the base of (TR10). Sync information, amplified, will be observed at the collector.

The sync limiter base is biased at −2.1 volts, for clamping and detection action. Clamping occurs on the top of the sync pulses so the average is negative. If no bias is measured, or the base is positive, then the transistor is open, or (C50) is open and not charging. The −2.1 volts at the base indicates normal sync limiter action.

When a noise pulse appears at the noise gate (TR11) base, and the stage moves out of saturation, sync action momentarily ceases for the duration of the pulse, but immediately recovers, as follows.

- A 0.22 μF capacitor (C50) is in series with the base emitter diode junction of the sync limiter transistor (TR10) to ground, under normal conditions of the noise gate (TR11) in saturation.
- Since the "diode action" of the transistor junction is a one-way gate, the capacitor C50 charges on composite information, creating the average negative value of −2.1 volts on the sync limiter base.
- The 820 k resistor, off sync limiter base to the 24 volt B+, properly biases the base so that conduction occurs in the middle of the sync tip information.
- When a noise pulse takes the noise gate (TR11) out of saturation, and the sync limiter emitter voltage increases, causing the stage to cease operation, the C50 capacitor loses its charge since composite video information is momentarily not conducting through it in the one-way path.
- After the noise pulse is passed, and the noise gate returns to saturation, the sync limiter, on the next horizontal sync pulse, immediately reestablishes normal operation.

Audio and Sound i-f (Solid-State B&W TV)

The solid-state audio system may be checked with a 0.1 μF capacitor from the power transformer on the CRT filament winding. This gives you a 60 Hz test injection signal for circuit troubleshooting.

Troubleshooting and Alignment

Remove the signal lead from the sound i-f module, as this eliminates background noise, and turn the volume up to full level. Refer to the i-f and audio circuits in Fig. 4-20.

A clip lead from the 0.1 μF test capacitor to the collector of the sound output transistor, (TR15), should produce a very weak 60 Hz hum.

At the base of TR15, a louder hum, should be detected if the audio output transistor is amplifying properly.

At the collector of sound driver (TR14), you should find the same hum level.

At the base of TR14, there should be a louder signal, and the volume control will have no effect.

At the lead from the i-f module, the same sound level as above should be heard, but the volume control will affect the sound output.

Now reconnect the sound i-f lead, if the audio amplifier circuit has been found to be good.

Sound Alignment

Remove the sound i-f and ratio detector module from the main chassis and keep the cover in place.

Now perform an alignment at 4.5 MHz. Insert 24 volts B+. Insert a 4.5 MHz carrier modulated with an audio tone.

Next check for response and audio limiting.

Apply 4.5 MHz unmodulated carrier at the input. Connect VTVM for dc measurement at audio i-f output lead, disconnected from the audio circuit. Adjust the frequency a maximum of ± 50 kHz either side of the 4.5 MHz center frequency. The meter should be observed to have linear swing of the output voltage with frequency change.

Shift the frequency, on a strong signal, so that the output voltage reads +1 volt dc. Now adjusting the input voltage from 100 mV down to 20 mV should produce no change in the meter reading. This tests limiting and sensitivity. If the 4.5 MHz carrier does not produce dc voltage swing, remove the case cover, which will detune the 4.5 MHz circuitry. Inspect the chassis for any obvious damage.

Check dc continuity at the output circuitry. Inspect T4, and the ratio detector output load.

See if the diodes are shorted. Check front to back ratio.

Connect 24 volt B+ and check for voltages as shown on the circuit diagram.

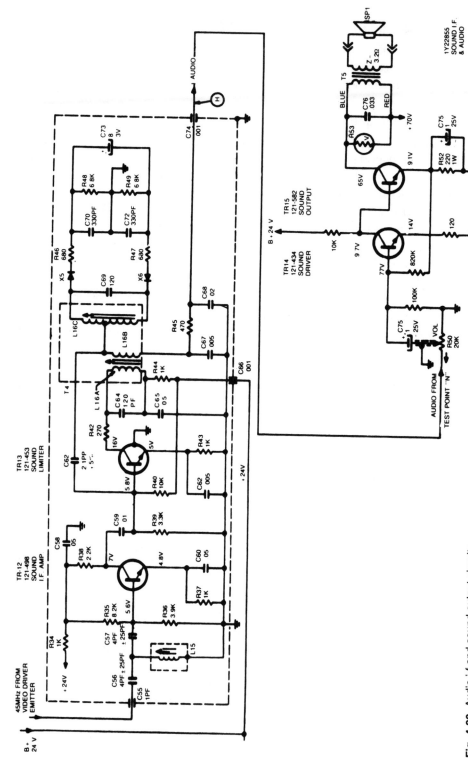

Fig. 4-20. Audio i-f and sound output circuits.

166

Circuit Operation

Briefly, limiting action in the ratio detector is accomplished partly by having equal dc voltage developed across each half of the L16C secondary transformer winding, across R48 and R49, and partly due to the electrolytic capacitor C73.

The 680 ohm resistors are primarily to offset variations in forward resistance of the matched diode pair (X5 and X6). Although the diode pairs are "matched" one may have a forward resistance of 20 ohms, the other somewhat different. However, the 680 ohms ensures that the detecting and limiting action is not impaired.

5

AM/FM Stereo
Tuners/Amplifiers
and TV Stereo Audio

In this chapter AM/FM stereo tuner and amplifier circuits' operation and troubleshooting will be explained. The last portion of this chapter will cover TV stereo theory of operation, troubleshooting tips, and some actual TV stereo audio circuit operation.

FM STEREO RECEIVER CIRCUIT (G.E. TU-100)

The General Electric TU100 FM stereo receiver circuit is shown in the semi-block diagram Fig. 5-1. The detected FM or FM stereo signal is fed to the base of transistor Q14, the composite signal amplifier. FM audio or composite FM stereo signals are amplified and appear in the collector circuit of Q14. The FM audio or composite FM stereo signals then pass through the band pass and 67 kHz filter to the stereo detector and de-emphasis circuits. Control R98 and capacitor C116 located in the emitter circuit of Q14 form the stereo balance control. Adjusting R98 establishes feedback to increase or decrease the gain in the L − R signal area.

Transformer T8B in the collector circuit of Q14 selects the 19 kHz pilot signal when it is present. The 19 kHz pilot signal is amplified by Q13. Control R96 and R95 form a series voltage divider network to supply base voltage to Q13. Thus, R96 acts as a threshold control by establishing the gain of Q13. The factory adjustment of R96 is to set the gain of Q13 for threshold at 6 percent of the pilot signal level. That is, Q13 will conduct and turn on Q11, thus lighting the FM stereo lamp indicator.

Transistor T9 in the collector circuit of Q13 is tuned to 38 kHz, the second harmonic of the 19 kHz pilot signal. The 38 kHz signal is amplified by the doubler stage Q12. T10B in the collector circuit of Q12 is inductively coupled to T10A, thus adding the regenerated 38 Hz subcarrier signal to the stereo detector circuit to detect the left and right FM stereo signal.

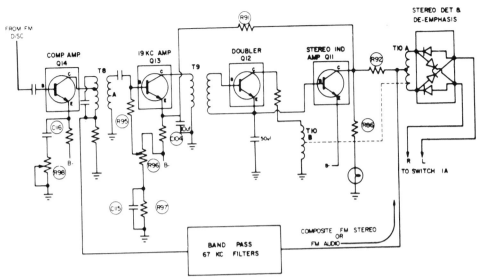

Fig. 5-1. Block diagram of stereo receiver.

When the 38 kHz regenerated signal is present at T9, capacitor C101 becomes charged and causes the dc level at the base of Q11 to go more positive, thus causing Q11 to conduct. Since the stereo indicator lamp is in series with the collector of Q11, it will light when Q11 is driven into conduction by the 38 kHz signal.

A portion of the collector voltage of Q11 is fed back to the emitter of Q13 via R91 to increase the forward bias and gain of Q13, firmly locking the circuit in tight. This prevents the FM stereo circuits from switching back and forth from stereo to monaural on weak or fading station signals.

AM/FM i-f AMPLIFIER CIRCUIT (G.E. TU-100 CHASSIS)

Transistors Q2 and Q3 serve as i-f amplifier stages for both AM and FM. The i-f coils are unusual in that they are capacity coupled as opposed to inductively coupled. The circuit for this i-f stage formed by T3 and T4 is shown in Fig. 5-2. Note that T3 consists of the i-f coil for FM and AM in the collector circuit of Q2, while T4 consists of the tuned coil for FM and AM in the input circuit of Q3. The top winding of each i-f coil is the FM section. T3 and T4 are located so that inductive coupling between coils is not possible. Note that the FM and AM coil windings of T3 are wound on one coil form, while the FM and AM coil windings of T4 are wound on another form. Both T3 and T4 are enclosed in a shield. The FM and AM i-f coils are slug tuned. Capacitor C18 couples the FM i-f signal from T3 to T4. C19 couples the AM i-f signal from T3 to T4.

With an FM i-f signal (10.7 MHz) present at T3, the following conditions exist. The bottom winding of T3, (AM section) acts as a choke or open circuit to the FM i-f frequency. This places C17, 820 pf and C16, 140 pf in series across the top winding (FM section) of T3.

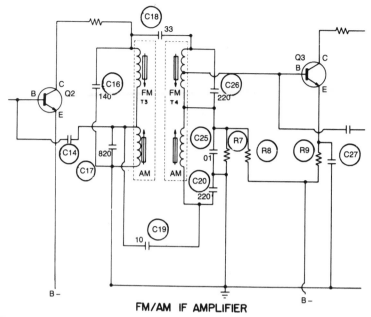

FM/AM IF AMPLIFIER

Fig. 5-2. FM/AM i-f amplifier circuit.

The two capacitors in series then act as the resonating capacitance for the top half of T3. C16 and C17 also act as a capacitive voltage divider from which C14 receives a neutralizing voltage to feed back to the base of Q2. C18 acts as a coupling capacitor to couple the FM i-f signal to the top winding of T4. The bottom winding (AM section) of T4 acts as a choke or open circuit to the FM i-f signal. Capacitor C26 acts as the resonator capacitor across the top winding (FM section) of T4. A capacitive voltage divider is formed by C25 and C26 so that the correct amount of ac signal is fed to the base of Q3 through the tap on the FM coil of T4.

When we consider an AM i-f (455 kHz) signal at the top of T3, the following conditions exist. The top winding (FM section) of T3 acts as short circuit to the AM i-f signal, thus C16 and C17 are in parallel across the AM winding (bottom half) of T3. Capacitive voltage division takes place in C16 and C17 to supply the neutralizing voltage source for C14 to feed back to the base of Q2. C19 acts as a coupling capacitor to feed the i-f signal to the AM (bottom winding) section of T4. Capacitors C20 and C25 are then in series across the AM (bottom winding) section of T4. Since the point between C20 and C25 is grounded, a capacitive voltage divider is formed so that the correct amount of AM signal is fed to the base of Q3 through the FM (top winding) section of T4. The FM (top winding) section of T4 acts as a short circuit to the AM i-f signal when coupling the AM i-f signal to the base of Q3. Capacitors in parallel with the i-f coil windings and coupling capacitors are mounted externally from the i-f coils as opposed to being an integral part of the i-f transformer base.

AM/FM TUNER POWER SUPPLY CIRCUIT

The voltage regulator circuit in Fig. 5-3 requires an input voltage of -22 to -35 volts dc to supply the AM/FM tuner with -12 volts dc $\pm 5\%$ of operating power. Voltage regulation is accomplished by zener diode D5, Q9, Q10 and associated components in a direct coupled circuit configuration. Transistor Q9 is forward biased by resistors R60 and R65 and zener diode D5. Transistor Q9 (PNP) is used as a voltage amplifier and polarity inverter to forward bias Q10 (NPN) into conduction and complete the voltage regulator circuit. Current limiting is performed by resistors R60, R65, R66, R67 and D5. Additional power supply filtering is accomplished by C86. Capacitors C87 and C88 filter any unwanted transient ac voltages that may be present in the input dc voltage and cause Q9 and Q10 to oscillate. Use a scope on base of Q9 to check for any oscillation that may be caused by an open filter capacitor(s).

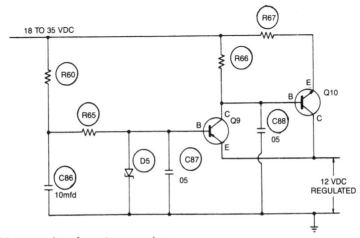

Fig. 5-3. Voltage regulator for a stereo receiver.

NOISE AMPLIFIER (SYLVANIA)

This noise amplifier and auto mono/stereo switching circuit is used in some Sylvania stereo tuner systems. In the FM multiplex signal, noise becomes more of an annoyance than in the FM mono signal, due to the AM characteristics of the multiplex signal. These AM characteristics permit noise to ride into the system, producing objectionable listening.

The noise amplifier transistor Q22, (see Fig. 5-4), is coupled to the FM detector chip, IC2, through a 330 pF capacitor, C124. When the noise level increases in the FM multiplex signal, the Q22 transistor amplifies the noise, and feeds it through C152 to diode SC4. The noise signal is rectified by SC4, producing a negative dc output. This voltage reduces the forward voltage applied to pin 4, IC4, by the resistance network, R126, R124, and SC4, causing Q31 to turn down. Its collector voltage rises, turning on the Darlington amplifier, Q29-Q30. The emitter voltage rises at Q29 turns on Q28, as its collector is connected to the junction of R14 and R16. It brings this junction point to ground.

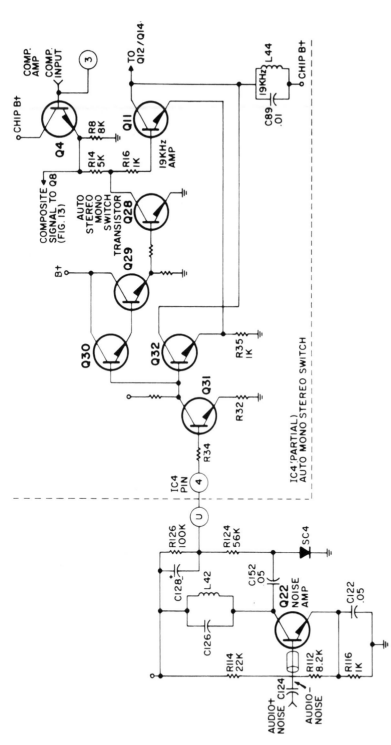

Fig. 5-4. Mono/Stereo IC4 audio switching circuit.

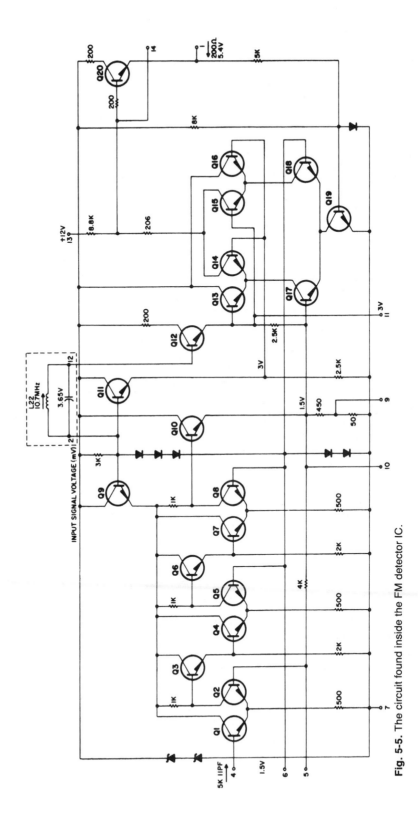

Fig. 5-5. The circuit found inside the FM detector IC.

173

The composite signal fed from Q4 to Q11 is shorted, killing the 19 and 38 kHz circuits. However, the composite fed from Q4 to Q8 drives the differential amplifier and synchronous detector through Q22. Without the 38 kHz switching signal, the decoder becomes an amplifier feeding the audio signal onto the stereo power amplifiers.

FM DETECTOR (IC2) SYLVANIA STEREO

This FM detector, IC2, shown in Fig. 5-5 has internal voltage regulation and good sensitivity. The FM detector chip has 20 transistors, 21 resistors, 6 diodes and two Zener diodes. This chip is an FM limiter, quadrature oscillator, and a product detector. FM limiting is done by three stages of differential amplifiers, using 9 transistors.

The quadrature oscillator Q12, and its external tuned circuit L22, receives the 10.7 MHz signal via a capacitive coupling in the L22 assembly.

Coil L22, adjusts both the phase and frequency of the 10.7 MHz quadrature oscillator signal and applies the correct phase to the bases of Q13, Q14, Q15, and Q16 in the product detectors. The FM i-f signal is applied to the emitters. The audio or composite multiplex signal is recovered by comparing the phase and frequency of the two 10.7 MHz signals across the emitter base junction of the product detector transistors.

i-f CIRCUIT STRIP—FM STEREO (ZENITH)

This i-f strip circuit consists of i-f transistor Q201, ceramic filter Y201 and IC201, which includes gain, limiter, and quadrature stages. Note FMIC circuit in Fig. 5-6.

As a reminder, ceramic filters are highly selective devices. Being made as fixed frequency devices, they may fall into one or more overlapping groups of frequencies. The center frequency of a ceramic filter may not be at the designated i-f frequency (such as 10.7 MHz). The filters may be grouped as shown in Table 5-1.

Table 5-1. Ceramic Filter Table.

TYPICAL CERAMIC FILTER GROUPS		
NOMINAL CENTER FREQUENCY	FREQUENCY RANGE	COLOR CODE
10.64 MHz	10.61 to 10.67 MHz	Black
10.67 MHz	10.64 to 10.70 MHz	Blue
10.70 MHz	10.67 to 10.73 MHz	Red
10.73 MHz	10.70 to 10.76 MHz	Orange
10.76 MHz	10.73 to 10.79 MHz	White

When aligning a ceramic filter, set the signal generator to the frequency of the ceramic filter, then align the i-f.

The FM-i-f, IC201 is preceeded by Q201 (first FM i-f) and ceramic filter Y201, with input to IC201 being at pin 1. This IC's internal circuitry (note block diagram in Fig. 5-7) consists of three i-f amplifier stages with level detectors for each stage, a doubly-balanced quadrature detector, an AFC amplifier, an audio amplifier and an internal voltage regulator.

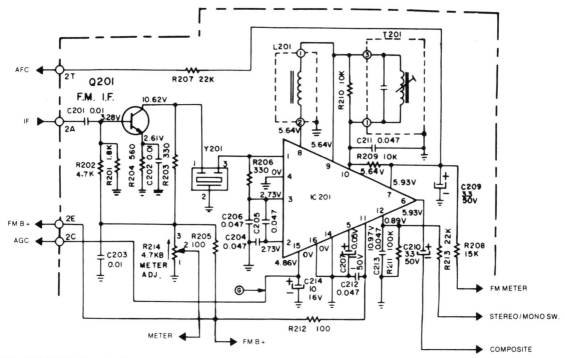

Fig. 5-6. FM i-f chip circuit.

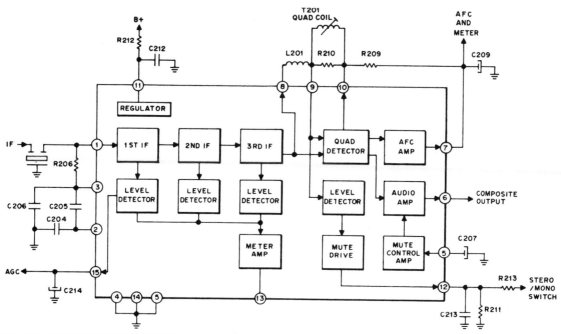

Fig. 5-7. Block diagram of FM i-f chip.

Outputs of IC201 include delayed AGC (pin 15), AFC and FM meter (pin 7), composite audio (pin 6) and mute control which is located at pin 12.

Delayed FM AGC voltage is developed internally within IC201. Output of the first i-f amplifier is amplified by its level detector. Since the first i-f amplifier is the last i-f amplifier to go into limiting, the level detector will develop a current which will appear as a voltage across C214, at IC201 pin 15. This is the delayed AGC voltage which is applied to the base of Q1 (FM-RF). When the i-f signal level is sufficient, Q1's base voltage is lowered, lowering the gain of transistor Q1.

FM-QUADRATURE DETECTOR

The amplified and limited i-f signal appears at IC201 pin 8 and is coupled via L201 to pin 9, the Quadrature detector. Connected externally, between pins 9 and 10 is Quad coil T201. Voltage appearing across T201 is a function of the signal frequencies appearing at the ends of T201. Signals at these pins will be 90 degrees apart, or in quadrature, resulting in the circuits being called a quad detector. Signals at the center frequency will produce equal voltages at pins 9 and 10, while frequencies off of center frequency will create different voltages on pins 9 and 10. Since this IC uses a differential circuit, only the difference of the signals appearing at pins 9 and 10 will be amplified. If pins 9 and 10 have identical signals, the signal will not be amplified.

Tuning Meter Operation

The tuning meter is common to both AM (maximum reading) and FM (zero center reading). Diode CR253 is the AM meter rectifier, while FM meter voltage comes from pin 7 of IC201 (FM-i-f). Bandswitch S1, selects the appropriate meter voltage. Meter adjustment resistor R214 is adjusted to set zero center for the meter as part of the FM alignment procedure.

Stereo/Mono Switching

Pin 12 of IC201 provides a control voltage which is used to control the stereo/mono switching of the stereo multiplex decoder, IC301.

FM MULTIPLEX CIRCUIT OPERATION (ZENITH)

A block diagram of IC301 is shown in Fig. 5-8 of the FM Multiplex Decoder. The circuit for the IC301 decoder is shown in Fig. 5-9. This decoder uses a voltage controlled oscillator (VCO) as part of a phase-locked loop (PLL). IC301 has three basic functions.

- Regeneration of the 38 kHz subcarrier frequency.
- Stereo indicator switching.
- Decoding (Matrixing L + R and L − R/38 kHz to provide L and R outputs).

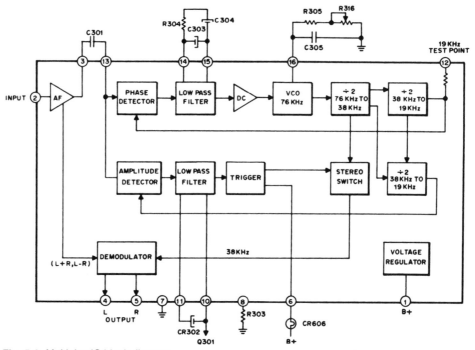

Fig. 5-8. Multiplex IC block diagram.

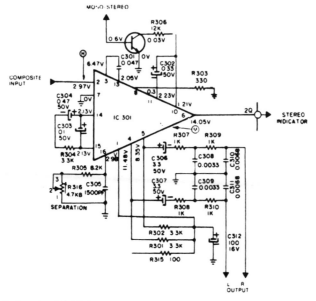

Fig. 5-9. FM Multiplex circuit (chip).

The composite signal (L + R, L − R, 38 kHz) from pin 6 of the FM-i-f (IC201) is connected to the input (pin 2) of IC301. This signal is amplified and then coupled from pin 3, via C301, to pin 13. It is also internally connected from the amplifier to the demodulator. From pin 13 the composite signal is divided into two paths (one is the phase detector and the other is an amplitude detector). Near the middle of the top row of blocks (at pin 16) is the voltage controlled oscillator (VCO) which runs at 76 kHz. Its free running frequency is determined by C305, R305 and R316. The VCO output goes to two divide-by-two stages, resulting in outputs of 39 kHz and 19 kHz. The 19 kHz is available at pin 12 for frequency measurement.

Proper adjustment of the VCO frequency is made by connecting a frequency counter to pin 12 of IC301 and adjusting R316 until the frequency counter reads 19 kHz. If you do not have a frequency counter, use the following technique. A station broadcasting a stereo program is tuned in, R316 is then adjusted until the stereo indicator turns on, followed by adjustment of R316 to the middle of the turn-on range.

The 19 kHz signal at pin 12 is fed back to the phase detector, where the phase and frequency of the 19 kHz signal is compared with the 19 kHz in the signal applied to pin 13. Any difference is fed to the low-pass filter (including external components) at pins 14 and 15. Output of the low-pass filter is converted to a dc correction voltage that is applied to the VCO to correct for any changes in the VCO frequency.

Also at pin 13 is the amplitude detector which receives both the input 19 kHz signal and the 19 kHz signal from a third divide-by-two stage. Incoming levels of the 19 kHz pilot level will be sensed by the amplitude detector. If the pilot level exceeds a minimum level, the amplitude detector will output a signal to the low-pass filter (including external components) at pins 10 and 11 and also to the trigger stage. When the trigger stage is turned on it will activate the stereo indicator, CR606, via pin 6 and also will enable the internal stereo switch circuitry to pass 38 kHz to the demodulator. In the demodulator, the sum (L + R) and difference (L − R) and 38 kHz signals are combined to derive both the "L" channel and the "R" channel.

Automatic stereo/mono switching is determined by the i-f signal level. This signal is converted to a control voltage which is available at pin 12 of the FM-i-f, IC201. Voltage on pin 12 will be high at zero i-f level but, as signal level increases, voltage at pin 12 drops as shown in Fig. 5-10. This voltage will be coupled, via R213, to Q301 (multiplex noise filter) where it will be amplified. When the i-f signal is low, a high voltage appears at IC201 pin 12, it is amplified by Q301 and is then applied, via R306, to pin 10 of the multiplex decoder, IC301. IC301 pin 10 is the output of the amplitude detector filter connected between pins 11 and 10. When a high control voltage is applied to pin 10, it will effectively switch off IC301's internal switching circuit, causing the circuit to revert to mono operation.

As the i-f signal level increases, voltage at IC201 pin 12 will drop, Q301 will cut off, voltage on IC301 pin 10 will go low and the incoming 19 kHz pilot signal will control the internal switching, passing stereo to IC301 output pins 4 and 5 of the chip.

A regulator is located within IC301 (connected at pin 1) and supplies most stages, while unregulated voltages power the other stages.

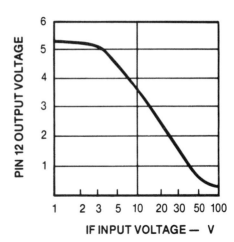

Fig. 5-10. Voltage chart.

Stereo Indicator Lamp (CR606) is connected to IC301 pin 6. When IC301 detects the presence of a stereo pilot signal, pin 6 will be enabled (go high) and a path will be completed through CR606 to a B+ voltage point on the power amplifier chassis.

Typical waveforms found in this FM Stereo Multiplex Decoder are shown for all active pin-outs in Fig. 5-11.

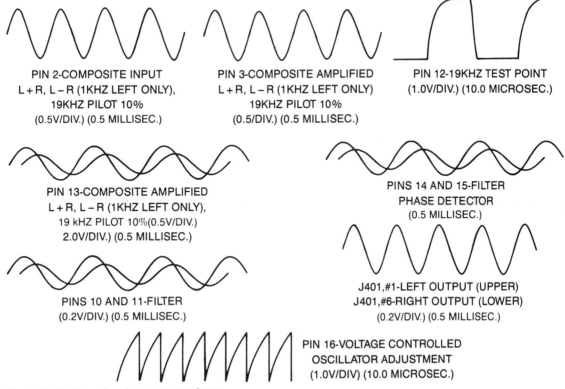

PIN 2-COMPOSITE INPUT
L + R, L – R (1KHZ LEFT ONLY),
19KHZ PILOT 10%
(0.5V/DIV.) (0.5 MILLISEC.)

PIN 3-COMPOSITE AMPLIFIED
L + R, L – R (1KHZ LEFT ONLY)
19KHZ PILOT 10%
(0.5/DIV.) (0.5 MILLISEC.)

PIN 12-19KHZ TEST POINT
(1.0V/DIV.) (10.0 MICROSEC.)

PIN 13-COMPOSITE AMPLIFIED
L + R, L – R (1KHZ LEFT ONLY),
19 kHZ PILOT 10%(0.5V/DIV.)
2.0V/DIV.) (0.5 MILLISEC.)

PINS 14 AND 15-FILTER
PHASE DETECTOR
(0.5 MILLISEC.)

PINS 10 AND 11-FILTER
(0.2V/DIV.) (0.5 MILLISEC.)

J401,#1-LEFT OUTPUT (UPPER)
J401,#6-RIGHT OUTPUT (LOWER)
(0.2V/DIV.) (0.5 MILLISEC.)

PIN 16-VOLTAGE CONTROLLED
OSCILLATOR ADJUSTMENT
(1.0V/DIV) (10.0 MICROSEC.)

Fig. 5-11. Multiplex decoder scope waveforms.

Multiplex Alignment Comments

Before any attempt is made to align, or service, FM multiplex circuitry, you must be sure that the RF, i-f, and detector alignment is correct, and that the receiver functions OK on monaural station signals.

Multiplex generators are good troubleshooting devices because they provide a composite multiplex signal as well as an RF signal (which is FM modulated by the composite multiplex signal). The composite signal is very useful since it can be used in signal tracing the multiplex portion of the receiver. It's not recommended that multiplex alignment be made using the composite signal injected at the output terminal of the detector since there is always some phase shift occurring in the RF, i-f or detector circuits. As a result, multiplex alignment made by a signal injected at the detector input would not be correct. For proper multiplex alignment, the composite signal must FM modulate the RF carrier and then be fed into the FM antenna input. With the signal injected in this manner, the multiplex alignment would be the best that could possibly be obtained.

RF signals should be injected at a point in the FM band where no signal is present. If at all possible, this should be at a frequency near the middle of the FM band. Tune the FM receiver to this point and adjust the RF frequency adjustment on the generator to this same frequency. The AGC voltage developed in the receiver should be maximum. An AGC voltage a lot less than this could indicate that the RF frequency is tuned to an image.

Stereo Tuner Troubleshooting Tips

Should a problem arise in aligning the FM multiplex portion of the receiver, you must determine whether the fault lies in the RF, i-f, and detector portions of the receiver, or whether the trouble is in the multiplex portion. The composite output of the multiplex generator can be injected at the output of the detector to help determine the area of difficulty. To reproduce possible extraneous signals coming through the detector, short the detector primary with a jumper lead. The waveforms and their amplitudes may vary slightly from chassis to chassis, however, they are quite indicative of what will be seen when signal tracing the multiplex circuitry.

If all of the waveforms are similar in form, it can be assumed that the multiplex portion of the receiver is functioning properly and the problem area is ahead of this in the FM receiver. If any of the scope waveforms are missing at a later point but are apparent at a previous point, circuitry between the two test points should be checked.

AUDIO AMPLIFIER CIRCUITRY (MONO/STEREO)

In the circuit shown in Fig. 5-12 the signal from the crystal cartridge is fed to the base of input transistor Q1 via the tone control. The signal goes onto the volume control, the isolating resistor R1 and coupling capacitor C2.

Q1 is a high Beta NPN silicon device operated class "A" and connected in the common collector mode. The stage has the following characteristics—high input imped-

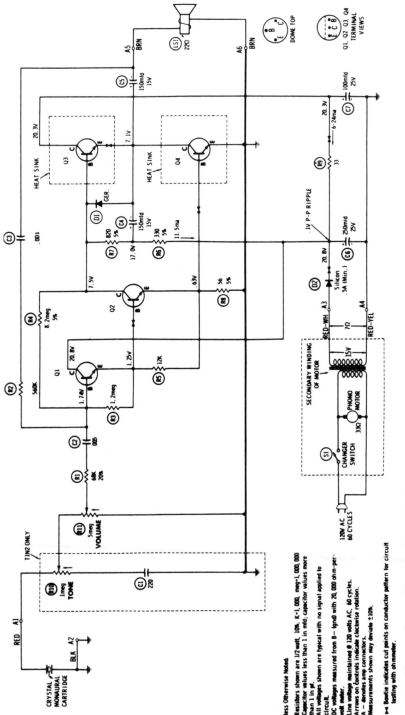

Unless Otherwise Noted:

1. Resistors shown are 1/2 watt, 10%. K=1,000, meg=1,000,000
2. Capacitor values less than 1 in mfd; capacitor values more than 1 in pf.
3. All voltages shown are typical with no signal applied to circuit.
4. DC voltages measured from B— (gnd with 20,000 ohm-per-volt meter.
5. Line voltage maintained @ 120 volts AC, 60 cycles.
6. Arrows on controls indicate clockwise rotation.
7. A — denotes amp connectors.
8. Measurements shown may deviate ±10%.

9. ⊶ Bowtie indicates cut points on conductor pattern for circuit testing with ohmmeter.

Fig. 5-12. Mono audio amplifier circuit (transistor).

181

ance—low output impedance—output signal has same phase as input-high current gain.

There is 7 db of feedback from the output fed to the base of Q1 via R2 and C3 to compensate for circuit nonlinearities. The output signal at the emitter of Q1 is directly coupled to the base of Q2. Transistor Q2 is a class "A" operated NPN silicon unit functioning as a phase splitter-driver. Q2 supplies driving signals for both output transistors. At its emitter, a current amplified signal in phase with that at its base is available for driving transistor Q4. At the collector of Q2, a phase inverted voltage and current amplified signal is used for driving Q3.

Q3 and Q4 are both NPN silicon units capable of 700 mW of power when heat-sinked. Both are forward biased by a positive base voltage. Since the signal fed to the base of Q3 is of the opposite phase from that at the base of Q4, forward bias is increased on one transistor by one half of the signal, while the other half increases the forward bias of the other. Thus, when class "B" operation is attained, each amplifies only one half of the signal. Q3 is connected as a common collector, thus, its output has the same phase as its base. Q3 provides power amplification of the positive half of the signal across the speaker load. Q4 is connected as a common emitter, and therefore, supplies a power amplified, phase inverted signal to the speaker load. In this way, Q4 provides an amplified negative going signal at its output load.

Two circuit components that play a very important part in this circuit's operation are capacitor C4 and diode D1. Their functions are as follows. C4 in series with C5 and the low impedance voice coil of the speaker act as a B+ filter and greatly reduce hum and ripple. C4 also provides a boot-strapping effect by supplying driving power to the collector of Q2, thus making possible wider signal swings before limiting takes place.

Diode D1, increases greatly the ability of this circuit to produce rated output power. D1 operates in the following manner. When Q2 conducts heavily with the application of a high amplitude signal at its base, its collector voltage drops rapidly due to current flow through the high series resistance of R7 and R6. When this potential drops slightly below the charge on C5, D1 becomes forward biased and allows the charge stored on capacitor C5 to furnish much needed driving power to transistor Q2.

The output before limiting of this amplifier is reduced to above 40% with this diode out of the circuit. Due to the use of dc feedback from the power supply, the operational characteristics of these amplifiers are quite unique. Due to the relatively high biasing of the power transistors in idle conditions, low amplitude signals are amplified class "A", thereby eliminating low level crossover distortion. As signal drive increases beyond class "A", operational power supply current increases. When this occurs, dc feedback reduces bias on the output transistors. Therefore, as signal drive increases the output transistors go from class "A" to class "AB" and finally to class "B" operation. Although crossover distortion does occur with class "B" operation, the amplitude of the sound when this point is reached is such that the distortion is masked out.

Normal dc (no signal) operation is usually indicated by an idling current of 6 to 24 ma for Q3 and Q4. The value of this current varies with circuit components, temperature and the ac line voltage. Very low line voltage reduces idling current and causes low level crossover distortion. High line voltage increases idling current. There should not be any detrimental effects even while operating the amplifier on line voltages up to 130 volts ac.

STEREO OUTPUT STAGES (IC AND TRANSISTOR)

Each channel of this stereo system contains one integrated circuit (IC1401), two driver transistors (one NPN Q1401 and one PNP Q1402) and two output transistors (one PNP Q1403 and one NPN Q1404). One circuit channel for this stereo system is shown in Fig. 5-13.

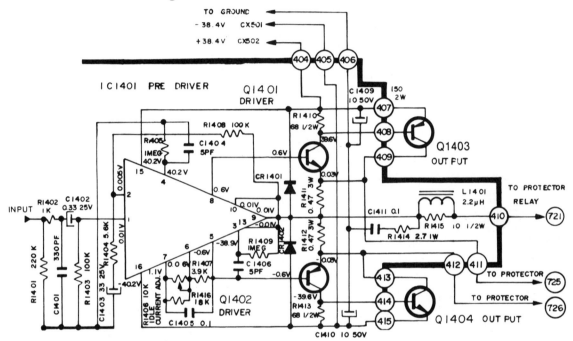

Fig. 5-13. Audio amplifier output circuit (IC).

Predriver IC1401 is a new high voltage device that operates from a split voltage supply, that is, both +38.4 and −38.4 volts are developed in the power supply to power these predriver IC's. These IC's contain differential amplifiers. Resistor R1406 at pins 6 and 7 provides for idle current adjustment. Outputs of IC1401 are at pin 8 (to driver Q1401) and pin 5 (to driver Q1402). Drivers Q1401 and Q1402 operate in push-pull, and drive the outputs of Q1403 and Q1404 respectively. Diodes CR1401, CR1402 protect against inductive spikes on the high voltage line, while L1401 is incorporated to compensate for capacitive loads at the output speaker terminals.

AUTOMATIC SPEAKER PROTECTION CIRCUIT

Special circuitry in the above stereo amplifier circuit provides automatic speaker protection in the event of either excessive voltage or excessive current. Note protection circuit shown in Fig. 5-14. In the event that a higher than normal voltage appears at

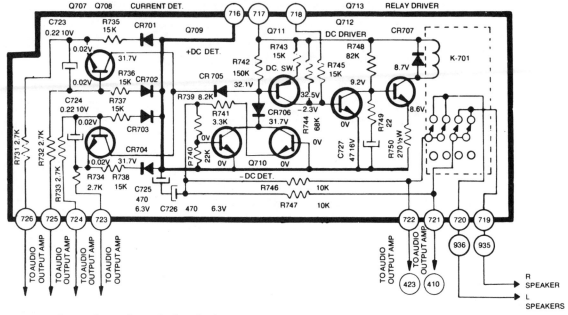

Fig. 5-14. Automatic speaker protection circuit.

the output (after L1401), it will appear across R747. If the voltage is in the positive direction, Q709 will conduct, turning off Q711 and Q712. This in turn switches Q713 off, deactivating relay K701. When the closed contacts on relay K701 are switched open, the circuit to the speaker connectors is then opened up. When the over voltage is in the negative direction Q710 will turn on, and in turn, Q711 and Q712 will switch on. This will also deactivate Q713 and relay K701, thereby protecting the speakers.

Current overload protection occurs when excessive current is sensed through R1411 and R1412. These currents will flow via R731 and R732. Small overloads will develop a voltage between base and emitter of Q707, causing Q707 to conduct. CR705 will be biased on, applying bias to Q711. Transistor Q711 will turn Q712 on and Q713 off, deactivating K701. When higher overload currents exist, they will cause a larger voltage across Q707's base emitter junction. This will cause diodes CR701 and CR702 to turn on, providing protection for transistor Q707.

POWER SUPPLY CIRCUIT

Two power supply sections are used for the amplifier circuit covered previously. These two power supply circuits are shown in Fig. 5-15. One supply is a full-wave bridge rectifier which is used to develop the split voltage required for the audio output stages. The second supply uses a full-wave rectifier with two regulator circuits. One regulator powers the RF/FM/multiplex circuits, while the second regulator (with voltage adjustment R509) powers the magnetic stereo cartridge and stereo preamplifier circuits that are used in another section of this system.

184

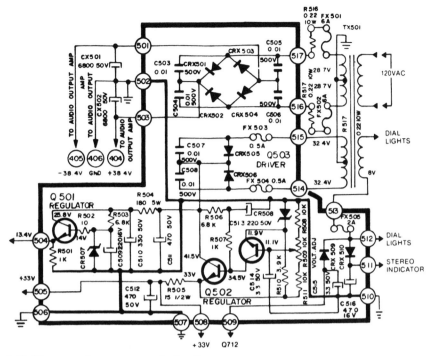

Fig. 5-15. Stereo receiver power supply circuit.

STEREO PUSH-PULL AMPLIFIER CIRCUITS

Stereo power amplifiers use push-pull arrangements primarily to minimize distortion. The push-pull circuit is most economical when high output power is required. Because the two halves of an ideal push-pull amplifier are perfectly balanced, the second harmonics are reduced and only the third and higher odd harmonics are of significance. This characteristic gives this type of circuit a major advantage in regard to overall distortion.

The low output impedance of transistors makes it possible to feed low-impedance loads directly. Because of this transistor characteristic, no matching device is required between the output transistor and the load, which is the speakers. Push-pull circuits can be classed as complementary and the quasi-complementary transistor amplifiers.

COMPLEMENTARY TRANSISTOR AMPLIFIERS

The complementary transistor is based on two elementary circuits. This is the Darlington and the complementary pair of transistors. Complementary amplifiers are produced when NPN and PNP transistors are used in series. Both the Darlington amplifier and the complementary pair are usable in individual push-pull circuits. These are shown in Figs. 5-16 and 5-17. In push-pull amplifiers, one output transistor conducts during the positive portion of the cycle and the second transistor conducts during the negative portion.

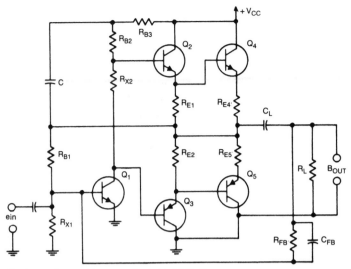

Fig. 5-16. Darlington pairs used in an audio output circuit.

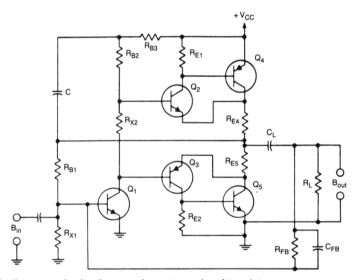

Fig. 5-17. Audio output circuit using complementary pairs of transistors.

In Figs. 5-16 and 5-17, both amplifiers are directly coupled to each other. In Fig. 5-16 the input signal is first amplified by Q1 and fed to the two Darlington pairs. The transistors in the upper Darlington pair are NPN. They will conduct when the signal at the output of Q1 is in the positive portion of the cycle. The PNP transistors in the lower Darlington circuit will conduct during the negative portion of the cycle. Both portions of the cycle are put back together as a complete cycle across the load resistor.

Complementary pairs have been substituted for the Darlington circuits in Fig. 5-17. Basically, amplifiers using both types of circuitry behave more or less the same. Voltage

186

for the bias circuit of transistors Q2 and Q3 is developed across RX2. Special diodes are frequently substituted for this resistor to improve bias stability, with temperature variations. A similar method is used for stabilizing a single-ended amplifier by the use of a diode in the bias circuit. In the collector of Q1, the load is the series combination of resistors, RX2, RB2 and RB3. One resistor, of a value equal to the two resistors, RB2 and RB3, would have done the job, but two resistors were used in order to connect a capacitor at their junctions. This capacitor is needed to overcome some circuit drawbacks.

Class B amplifiers, and to a lesser degree Class AB amplifiers, are prone to have distortion. Push-pull circuits are usually operated Class AB to reduce crossover distortion. In theory, when the upper transistor of Fig. 5-16 stops conducting in the class B mode of operation, the lower devices start conducting. However, there can be an instant in the cycle when the output transistors are not conducting. This characteristic causes what is known as crossover distortion. Considerable negative feedback is generally required in the circuit to eliminate or minimize the distortion. Negative feedback is accomplished by having the RFB-CFB network feed the signal back from the output to the input circuit.

In order to accommodate large amounts of feedback, the amplifier must have high gain. Capacitor C provides positive feedback in what is referred to as a bootstrap circuit. Positive feedback increases the input impedance of Q2 so that the load in the collector of Q1 is large. This large collector resistance is necessary if Q1 is to have high voltage gain.

This capacitor, C, serves a second very important function. Should the positive peak in the signal be large, it places the base and emitter of Q2 at the same +Vcc potential. Transistor Q2 then will be cut off at the peak in the signal. C comes to the rescue by being charged when the circuit is idling. The voltage developed across C will maintain the base positive with respect to the emitter at all times, even during positive peaks in the cycle. The transistor Q2, consequently, will not be turned off.

STEREO AMPLIFIER TROUBLESHOOTING

You may find that the troubleshooting of stereo amplifiers and FM multiplex systems can be greatly simplified by using a dual-trace oscilloscope and a square-wave audio or function generator.

Brief Procedure Tips

The basic concept of this test technique is to compare the normal operative channel with the inoperative stereo channel. Thus, one channel has to be operating properly. The square-wave audio signal is fed into both channels and the dual-trace scope is used to check and compare each channel's test points throughout all of the stages until the fault is located.

Of course, initial tests are performed for such obvious defects as excessive current drain, overheated solid-state devices, burnt resistors, poor solder connections, and cracked or broken circuit boards. Generally, after the preceding tests are performed and any faults corrected, at least one channel of the stereo system will be operating.

Stage Gain Checks

Since the scope's vertical amplifier gain controls on triggered-sweep, dual-trace instruments are calibrated in volts per centimeter, and have matched vertical channels, it's very easy to check from one test point to another, to compare the signal gain of each stage, and make a comparison check of the right and left stereo channels. With this technique you can quickly isolate the amplifier trouble in the defective stage and almost to the very component.

As a final check of the amplifier, or if the original symptom was for insufficient frequency response, the function generators audio test frequency can be tuned throughout the entire audio range for a response check. Make these checks at 5 kHz intervals and look at the dual-trace scope patterns for both speakers at the same time.

Intermittent Trouble Location Hints

This technique will help you locate those pesky intermittent troubles in a stereo amplifier. In some cases, the volume level may go up and down intermittently, go out completely, or it may operate for hours before acting up at all. The method I have found useful is to inject a square-wave signal into the suspected channel and connect both scope probes to different stages of the amplifier. When the volume fluctuates or some noise occurs, just glance at the scope screen and take note of any changes in the waveshapes. If there is no change in the pattern, move the probes to different stages and make some more checks. With this technique you can isolate the defective stage or component quite rapidly.

THE STEREO TV SYSTEM

To properly understand the TV stereo operation it is important to be familiar with the MTS (Multi-Channel TV Sound) system. The drawing in (Fig. 5-18) illustrates the spectrial diagram of the MTS signal.

The L + R portion or main channel of the MTS signal is identical to the current mono signal. The L + R signal has 75 microseconds pre-emphasis and deviates the aural carrier 25 kHz. The next component of the MTS signal is the 15.734 kHz stereo pilot. The pilot deviates the aural carrier 5 kHz and is used in the receiver to detect the L − R stereo subchannel. The L − R subchannel is an AM double sideband suppressed carrier signal and deviates the aural carrier 50 kHz. The SAP (Second Audio Program) subcarrier is an FM, 10 kHZ deviated signal centered at 78.670 kHz. The SAP subcarrier deviates the aural carrier 15 kHz. Both the L − R and SAP audio signals are dbx encoded to reduce noise and buzz.

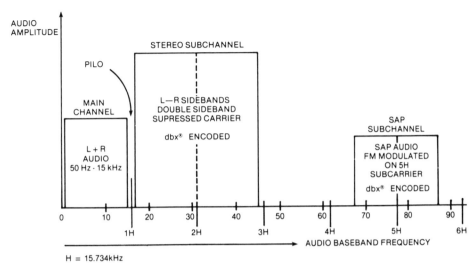

H = 15.734kHz

Fig. 5-18. Each of the signals making up composite audio occupies a different range of frequencies so that the receiver is able to separate and process each signal individually. (Courtesy of Sencore, Inc.)

Key Points

Now let's look at some of the key points found in the TV stereo sound system.

- The transmitter matrixes (mixes) left and right audio signals to form L + R and L − R signals. Refer to (Fig. 5-20).
- The main channel L + R signal is the sum of the two audio signals, it represents the similarity of the signals, and occupies the frequency range of 50 Hz to 15 kHz.
- The difference channel L − R signal occupying the same frequency range represents the difference between the two audio signals. The L − R signal is encoded in a dbx circuit to reduce the effects of noise generated in the MTS system. The encoded L − R signal is raised in frequency by amplitude modulating a 31.468 kHz (2H) subcarrier in a balanced modulator. The resulting stereo subchannel now occupies the frequency range of 16.468 to 46.468 kHz.

The stereo pilot, required by the receiver as a reference, is generated at 15.734 kHz. It is used to phase lock the receiver to demodulate the stereo subchannel signal and light the stereo indicator light.

The second audio program (SAP) signal is often included in the transmitted audio ranges from 50 Hz to 10 kHz. This signal is also encoded by a dbx circuit. The encoded SAP audio signal frequency modulates a 78.670 kHz (5H) subcarrier. The SAP signal is converted to an FM carrier and its sidebands that occupy the frequency range of 68.670 to 88.670 kHz.

The audio components are combined to form composite audio. Since the audio components (L + R, L − R, Pilot, and SAP) occupy different frequency ranges (50 Hz

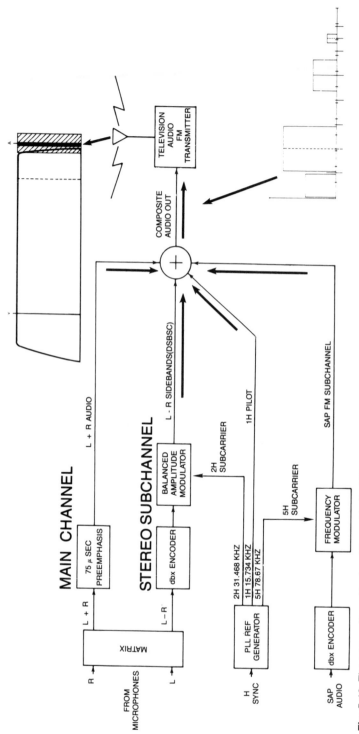

Fig. 5-19. The composite audio signal formed in the TV station audio circuits frequency modulates the audio transmitter. (Courtesy of Sencore, Inc.)

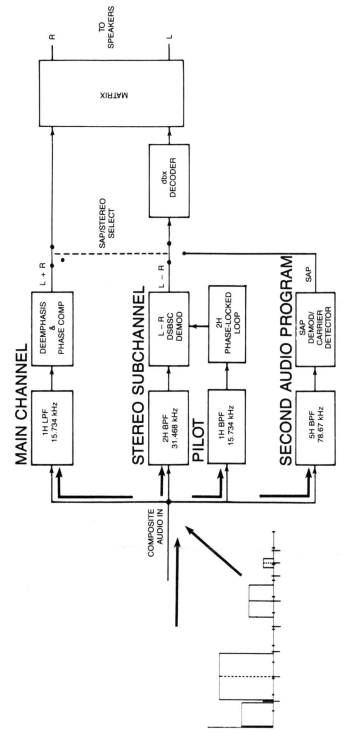

Fig. 5-20. The MTS receiver separates the composite audio signal to process each part of the signal in its own circuit. (Courtesy of Sencore, Inc.)

to 88.670 kHz) they do not interfere with each other. Refer to the block drawing shown in (Fig. 5-19).

The composite audio signal modulates the audio FM transmitter to form an audio carrier with frequency modulated sidebands. After transmission, the stereo TV receives this signal.

The stereo TV receiver splits the audio signals back up by acting upon the signal with a reverse process to recover left and right stereo signals or monaural SAP audio. Refer to the block diagram in (Fig. 5-20). Each part of the composite audio signal is treated differently to restore the individual signals to their original condition.

TROUBLESHOOTING TECHNIQUES

Let's assume a customer brings a stereo TV into the shop complaining of "poor stereo sound."

First, you want to verify that there is really a problem with the receiver. Some customers are served by a cable system that does not carry the stereo signal. In that case, the customer has standard sound on stereo broadcasts, although his receiver is capable of receiving stereo. The customer believes his receiver is not performing properly. You want to be certain a problem really exists in a stereo TV before probing around the circuits and chasing phantom problems.

Second, I know you want to quickly isolate the defective stage when there is a problem. The customer's complaint of poor stereo sound could be caused by any of the stereo TV stages between the antenna input and the speakers. You want to quickly zero in on the defective stage to avoid prolonged troubleshooting time and make a profit.

PERFORMANCE STEREO TV TEST

How do you know the stereo TV is defective? When you suspect that a stereo TV is operating improperly, or when your customer doubts his stereo operation, the first thing to do is performance test the receiver.

What better way to quickly performance test a stereo TV than right from the antenna input? This tests the television under actual operating conditions and saves the time otherwise spent finding test points within the set.

Apply a known signal to the antenna terminals and observe the results on the stereo TV, you will not waste your valuable service time trying to find phantom problems. When the performance test indicates proper stereo TV operation you will be confident telling your customer that his set is operating to the specifications set by the manufacturer of the TV. You may also make a reasonable charge for this service.

When the performance test indicates a problem with the stereo TV, you are certain that you're on track to a profitable repair. Let's now run through a "performance test" for a TV stereo system.

Two main signals are normally present at the Stereo TV antenna input:

1. An RF video carrier with composite video modulation:

- The composite video is made up of horizontal and vertical sync and blanking, video, and chroma.
- The level of the carrier is typically 1 to 3 mV.

2. An RF aural carrier with MTS composite audio modulation.

- The composite audio is made up of the L + R, L − R, pilot, and SAP signals mentioned earlier.
- The MTS stereo and SAP subcarriers are phaselocked to the horizontal sync signal.

While supplying (injecting) the RF aural carrier to the antenna input, you will also inject the RF video carrier to the antenna input with modulation appropriate to the test you are performing (usually a black screen for least video noise in the audio).

Note: You will need a special generator setup here to phase lock the MTS audio subcarriers to the horizontal sync.

Choose a channel free from strong local broadcasts to avoid interference problems. The RF level should be 1 to 3 mV. The audio carrier must be modulated by one or more of the standard MTS signal components mentioned earlier (L + R, L − R, pilot, SAP).

The generator's audio carrier must be modulated to a high enough level to avoid the erratic effects of noise. A 50% modulation level of the equivalent L − R channel accomplishes this and accurately simulates the broadcast station's transmission.

You will need to control these signals separately to test different stereo TV functions. Let's look at the performance tests you will want to perform at this test point—including the signals used.

HOW TO IDENTIFY STEREO TV PROBLEMS

During these tests, monitor the stereo TV audio outputs with a dual-trace scope, or better yet, use a dual-channel stereo monitor to verify the audio output signals. Sencore recommends the SC61 waveform analyzer or the new SM81 universal stereo monitor for fast, accurate measurements. Monitor the outputs at the line level output jacks on the back panel of the receiver or at the speaker output leads.

Let's look at the ten easy tests that are required to completely check out the TV stereo system.

One: Stereo Operation

To test for normal stereo operation, inject an RF aural carrier modulated with stereo signals to test point 1, as shown in the Fig. 5-21 block diagram. Use a medium frequency (1 kHz) tone to modulate the RF generator.

The stereo indicator lamp should light, both audio channels should produce a clean 1 kHz audio output, and there should be equal output from both channels.

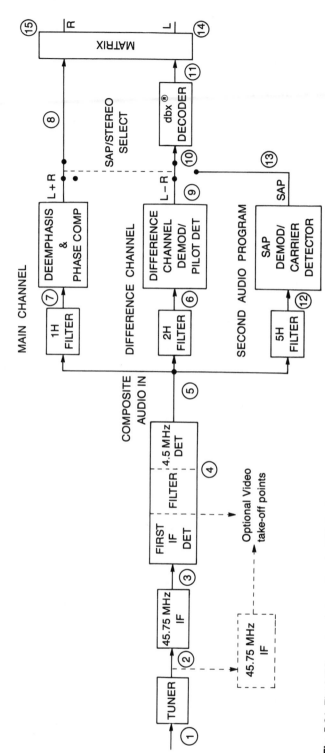

Fig. 5-21. The MTS receiver is based on the TV circuits familiar to all TV technicians. (Courtesy of Sencore, Inc.)

194

Two: Stereo Separation

Separation is the ability of the receiver to provide an output on one channel only, without a signal leaking to the other channel.

To test for separation, inject an RF aural carrier with single channel MTS stereo modulation (either right or left channel only) to test point 1. Use a low frequency (300 Hz) tone to check separation. Measure the output voltage of both the right and left channel outputs for a quick indication of separation.

Repeat the test with modulation on the other channel.

At least 18 dB (or the manufacturer's specs when available) for both tests indicates good separation.

Three: Stereo Signal-to-Noise Ratio

Perform this test to determine how much internal noise is being generated in the receiver stereo decoder circuitry.

Inject an RF aural carrier modulated with a stereo pilot only (without audio modulation). This enables the stereo circuits, but produces no audio output. Measure and record the higher of the two noise output voltages.

Modulate the RF signal with the stereo audio and pilot; use a low frequency (300 Hz) tone. Record the higher of the two audio output voltages and calculate the signal-to-noise ratio. At least 33 dB indicates good stereo channel noise performance.

Four: Stereo Pilot Switching Threshold

This test is important to insure that the stereo circuits switch on only when a stereo signal is really present and that they switch off when there is no stereo signal being received.

Inject an RF aural carrier modulated with only a stereo pilot. Decrease the pilot level until the stereo indicator just goes out. The switch-off point should be a least 50% of normal pilot level or lower (normal pilot level is +/− 5 Hz deviation of the audio carrier, you will need to set your generator for 2.5 kHz deviation for this test.)

Now increase the pilot level until the stereo indicator just comes on. The switch-on point should be at 50% of normal pilot level or higher.

Five: SAP-to-Stereo Cross talk

When there is signal leakage (cross talk) between circuits in the stereo receiver, SAP audio is heard during quiet moments of stereo operation.

To test for this cross talk, inject an RF aural carrier modulated with a stereo pilot only, without audio modulation. Measure and record the higher of the two noise output voltages. (This establishes a normal noise level for the stereo channel.)

Inject an RF aural carrier modulated with the stereo pilot and SAP signal with 1 kHz modulation (any cross talk increases the stereo channel noise level). Again, measure and record the higher of the two output voltages. Any increase in the voltage level indicates the presence of cross talk.

Six: SAP Operation

To test for normal SAP operation, inject an RF aural carrier modulated with the SAP audio signal to test point 1. Use a medium frequency (1 kHz) SAP modulation. The SAP lamp should light and a signal should be present at both audio outputs.

Seven: SAP Signal-to-Noise Ratio

Perform this test to determine how much internal noise is being generated in the receiver SAP decoder circuitry.

Inject an RF aural carrier modulated with a SAP carrier only without SAP audio modulation. This enables the SAP circuits but gives only noise at the output. Measure and record the higher of the two noise output voltages. Apply an RF aural carrier with SAP modulation (300 Hz). Record the higher of the two audio output voltages. At least 33 dB (or the manufacturer's specs when available) indicates good SAP channel noise performance.

Eight: SAP Carrier Switching Threshold

This test is important to insure that the SAP circuits switch on only when a SAP signal is really present, and that they switch off when there is no SAP signal received.

Inject an RF aural carrier modulated with the SAP carrier only. Decrease the SAP carrier level until the SAP indicator just goes out. The switch-off point should be at 50% of normal carrier level or lower (normal SAP carrier level is +/− 15 kHz deviation of the main audio carrier). Increase the SAP carrier level until the SAP indicator just comes back on. The switch-on point should be 50% of normal carrier level or higher.

Nine: Stereo to SAP Cross talk

When there is cross talk between circuits in the stereo receiver, stereo audio is heard during quiet moments of SAP operation. To test this cross talk, inject an RF aural carrier modulated with the SAP carrier only. Measure and record the higher of the two noise output voltages. (This establishes a normal noise level for the SAP channel).

Inject an RF aural carrier modulated with the SAP carrier and stereo signal with 1 kHz modulation (any cross talk increases the SAP channel noise level). Again, measure and record the higher of the two output voltages. Any increase in voltage level indicates cross talk.

Ten: Audio Beat Interference in Video

When the traps in the i-f are improperly adjusted, or when the i-f filter is defective, components of the audio signal cause interference in the video.

To test for this interference, inject an RF video carrier modulated with a video pattern to the antenna input. Also, inject an RF aural carrier modulated with each of the MTS audio modulations' components, one at a time. Watch the picture for signs of interference.

196

Performance Testing

Each of the ten special tests outlined is quick and easy to do if you have the equipment designed to supply all the necessary signals to the antenna terminals. Each test tells you immediately whether the performance is good and whether you should proceed with some troubleshooting checks. When you do have to troubleshoot, you will want all of the special signals used in MTS stereo TV at your fingertips.

6
Tape Deck Electronics and CD Players

In this chapter we will cover circuit operation and service tips on the electronic systems used in cassette and 8-track tape recorders. This will also cover stereo preamps, oscillators, auto-tape-stop circuits and bias erase circuits.

Recording a tape starts with an audio information signal or the intelligence to be recorded. To do this, a signal current is passed through a coil in a recording head. Tape, with a magnetic coating, is fed past the poles of the coils in the record head. The signal is recorded on the tape as magnetic fluctuations. The degree of these fluctuations is in proportion to the magnitude of the variations in the magnetic field.

For playback, the tape is drawn past another, or the same head in the machine. Magnetic variations on the tape induce current fluctuations in the head. These currents are amplified in the recorder circuits and reproduced as originally recorded on the tape. Because the frequency response of the playback amplifiers are not linear, some compensation circuits must be used for equalization.

A typical recording circuit is shown in Fig. 6-1. Audio from a microphone is fed to the gate of the n-channel MOSFET, Q1, preamplifier. Because the output from the microphone is linear (flat response), the gain of Q1 must also be uniform over the frequency range. The output from the MOSFET is developed across control R1.

A higher level signal (from turntable, tuner, or other tape unit) is fed to the line input across potentiometer R2. These input line signals are mixed via isolating resistors R3 and R4 and fed to bipolar transistor Q2. Transistors Q2 and Q3 serve the dual function of amplifying the signal while providing the circuitry to produce the proper frequency equalization.

Equalization or curve matching is accomplished through the use of feedback. Linear feedback takes place across resistor R5, but capacitor C4 causes the feedback to de-

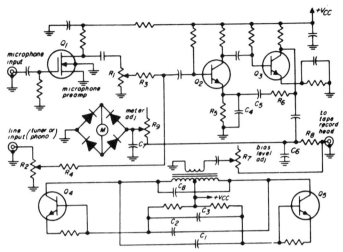

Fig. 6-1. Tape recorder preamplifier circuit. (Courtesy of Zenith Corp.)

crease as the frequency increases. Because gain is not reduced by feedback at the high frequencies, the needed treble is provided.

Resistor R6 and capacitor C5 feed signal back from the output of Q3 to the emitter of Q2. Because the impedance of C5 decreases as frequency increases, the high frequencies are fed back more readily than the lows. Thus, the low frequencies are emphasized here. This base boost is required to compensate some for the characteristics of the record head and the tape. The output from Q3 is fed to the record head.

Transistors Q4 and Q5 are part of the high-frequency multivibrator oscillator circuit. Capacitor C8, across the transformer primary of this circuit, sets the resonant frequency of the tank circuit and the frequency of oscillation. Capacitor C3 reduces the distortion of the sinusoidal output of the oscillator. Both the signal from Q3 and the high supersonic frequency are fed to the record head. The high frequency signal biases the record head to a specific level so that the signal rides on the oscillation. Because the bias level affects the frequency response, a bias level adjustment is put into this circuit to set the flow of high-frequency current through the head. The meter in the output circuit measures the record signal level. Control R9 calibrates the meter. Resistor R8 and capacitor C6 filter the high frequency so that it will not be fed back to the tape amplifier. It also helps in keeping the bias current out of the meter circuit. Capacitor C7 enhances the filtering action in the meter circuit.

THE PLAYBACK SYSTEM

In the playback mode the tape preamplifier must have the correct frequency characteristics. Refer to the playback circuit shown in Fig. 6-1. Transistor Q1 provides the high impedance that should be "seen" by the playback head and also supplies a constant gain over the entire audio bandwidth. The output level is varied by control R1. The output from R1 is fed to amplifier Q2 and Q3. The output signal from Q3 may be used to drive one of the power amplifiers in the tape system.

Parallel circuit R2-C1, in Fig. 6-2, furnishes some treble boost to compensate for the high-frequency roll-off of the playback head. R3-C2 provides the bass boost to simulate the playback curve. The action of both feedback circuits, discussed with respect to the record amplifier in the Fig. 6-1 circuit, also applies here.

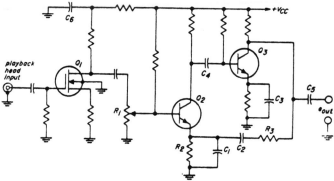

Fig. 6-2. Tape playback preamplifier circuit. (Courtesy of Zenith Corp.)

The inductance of the playback head must also be considered when measuring the playback characteristics. Playback curve requirements must be satisfied for the entire circuit. In the setup for measuring the frequency characteristics, the playback head must be placed between the audio test generator and the input to the playback preamplifier in series with the signal. The frequency response of the tape-head preamplifier combination must produce a curve that will give a linear output across the audio spectrum.

STEREO TAPE SYSTEM

Each channel of a stereo system has one record preamplifier and one playback preamplifier. Many good, high quality decks use two heads, an erase head and a single play/record head. The better machines use three heads, an erase head, a record head, and a playback head. Play/record heads, used on two-head machines, must be a compromise to perform both record and playback functions satisfactorily. The three-head decks are sufficiently versatile so that each head offers optimum performance for its particular function.

Each head intended for stereo operation has two individual coils. Two tracks (one for the left and the other for the right channel) are recorded when the tape travels in one direction from the supply reel to the takeup reel, and two other tracks are recorded when the tape travels in the other direction. On 8-track and some cassette recorders more tracks are recorded.

As shown in Fig. 6-3, tracks 1 and 3 are recorded when the tape travels in one direction, and tracks 2 and 4 are recorded when the tape is turned over and the functions of the two reels are interchanged. Two coils are necessary to record four tracks on the tape. The two channels recorded on two of the four tracks require two individual playback amplifiers.

200

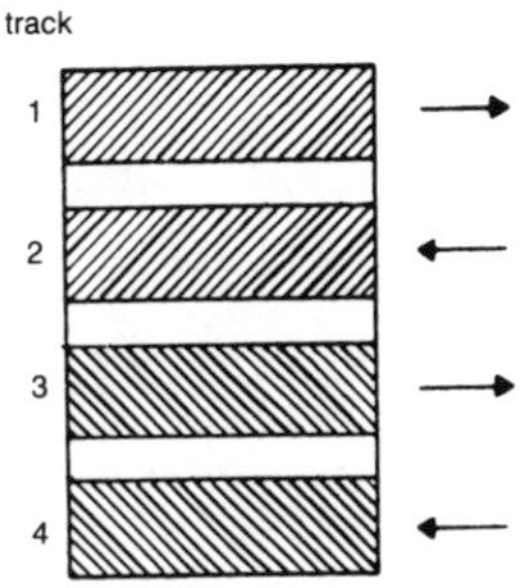

Fig. 6-3. Directions the tape tracks are recorded. Courtesy of Zenith Corp.

Many of the newer stereo tape recorders are now very versatile. You can use them to make sound-with-sound recordings on machines using two heads. As an example, it allows you to vocalize on one track while listening to music that was previously recorded on the other track. And of course you can also add (sound-over-sound) on both tracks. The circuitry for sound-with-sound operation operates as follows. One track on the head is switched so that it will play through its amplifier while the second track on the head is connected to the record amplifier. You can listen to music from the track feeding the playback amplifier while recording on the track connected to the record amplifier and to its respective coil on the head.

Additional versatility is possible if the stereo recorder has three heads. Music recorded on one track, possibly the left channel, is fed from the playback amplifier to the line input on the right-channel record amplifier. A live signal is simultaneously routed to the microphone input on the right channel preamplifier. Both signals at the right channels are then mixed. The composite signal is then recorded on one-right channel track. This recording method is known as sound-on-sound, and is achieved by connecting the output from the left-channel playback preamplifier to the line input on the right-channel record preamplifier. The left channel is shorted out at the record head. The right channel record and playback amplifiers are connected in the normal way to the right-channel coils of the respective record and playback tape recorder heads.

GENERAL CASSETTE TAPE RECORDER DATA

Stereo cassette tape players/recorders are used to play back prerecorded tapes and also to record from microphones or other sources such as AM or FM tuners, phono or Aux. Many cassette tape units feature playback and record, fast forward, play, stop-eject, pause and automatic tape shutoff. Some deluxe units will have tape select, VU meters, record level controls, digital counter and record indicator light.

Cassette Tape Specifications

The standard cassette tape width is 0.150 inches (slightly more than 1/8 inch) and is used for both monaural and stereo recordings. Figure 6-4 illustrates the track arrangements for these cassette tapes. As shown, a mono recording consists of two tracks; each track is 0.059 inch wide with a guard band of 0.032 inch between the tracks. Stereo recordings consist of two pairs of tracks. Each stereo track is 0.024 inch wide, sepa-

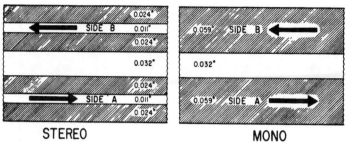

STEREO   MONO

Fig. 6-4. Cassette track patterns. (Courtesy of Zenith Corp.)

rated by a guard band of 0.011 inch. In addition, a guard band 0.032 inch wide separates each pair of tracks, thereby ensuring playback compatibility of the stereo recorded tape.

The total track width of each stereo pair, plus their guard band, is equal to one mono track width. Stereo prerecorded tapes will be reproduced in mono, on a mono tape unit. The left and right track signals will be combined by the playback head and be reproduced as a mono program. Recordings which are made on a mono unit will be reproduced only as mono, even on a stereo recorder system.

Cassette Electronics Circuitry

This cassette tape player/recorder uses an independent tape amplifier circuit board, with inputs and outputs connected to the systems other circuits. Other electrical connections to the main systems PC board are made directly or through switching networks.

The drive motor for this player/recorder is powered by 12 volts supplied from the main circuit board.

Refer to Fig. 6-5 for the electronic circuits used with this cassette tape player/recorder.

PLAY/RECORD AMPLIFIER OPERATION

This tape player/recorder uses four transistors and two integrated circuits (IC's). Each channel uses an IC for record/play amplification and equalization and one transistor for additional equalization. Also, one transistor functions as the bias oscillator while another transistor is used for B+ filtering.

A frequency compensating network provides the equalization required for proper record/playback response of the tape.

The S1417 switch (tape select) controls two different networks, one is used with chromium dioxide (CrO_2) tapes, and the other one is used with normal (NOR) ferric oxide tapes.

The amplifier circuit board utilizes a transistor in the bias oscillator circuit. Two controls are used to adjust the bias levels. Potentiometer R1701 controls the right channel bias level while potentiometer R1751 controls the left channel bias level. In

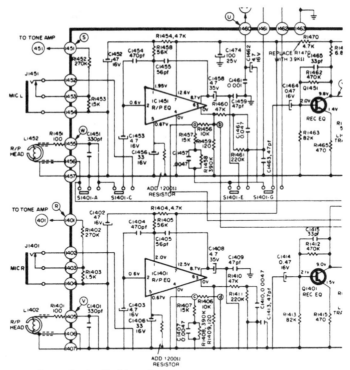

Fig. 6-5. Cassette electronic circuits (stereo). (Courtesy of Zenith Corp.)

addition, the unit uses record level controls (R1421) and (R1471) for both right and left channels.

BIAS OSCILLATOR CIRCUIT

The bias oscillator circuitry uses a transistor (Q1701) and a bias oscillator transformer (T1701). This circuit supplies erase current to the erase head L1701 and also supplies bias current to the play/record head. Bias current goes via recording bias level adjustments R1701 and R1751. This bias current is combined with the audio output signal after which it is applied to the respective left and right windings of the play/record head.

If the bias oscillator is not operating the tape cannot be recorded properly nor played back. Use the scope to check for bias oscillator operation. The bias oscillator produces a sinewave signal. A defective play/record head may also cause the bias signal to be very low in amplitude. One quick way to check for presence of the bias oscillator signal, without going into the tape unit is to connect another play head to the oscilloscope vertical input channel, and place the test head close to the play/record head in the tape player, and see if the sinewave bias oscillator signal is present on the scope.

Most of the bias oscillator units are sealed and cannot be repaired. If the unit is found to be defective just replace the complete bias oscillator unit.

PROFESSIONAL CASSETTE PLAY/RECORD TAPE UNIT ZENITH MC9070

A block diagram for this model MC9070 tape system is shown in Fig. 6-6. The signal source inputs are shown at the upper left of the diagram. The inputs are as follows.

Line input.
Microphone jack.
Dual-function record/play head (L1).
On the right side of the block diagram are shown these inputs:
Line output.
Headphone jack.

A level meter (M1) plus peak and recorder indicators are shown at the upper right corner, while the erase head (L2) is at the lower left.

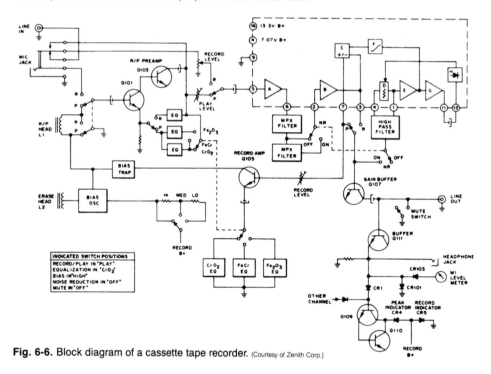

Fig. 6-6. Block diagram of a cassette tape recorder. (Courtesy of Zenith Corp.)

Cassette Play Mode

When in the PLAY position (with Dolby noise reduction turned off), the signal moves from the record/play head via one section of the record/play switch, to direct-coupled amplifier Q101 and Q103. Another section of the record/play switch selects the desired feedback circuit to provide proper equalization during record or playback. Output of this stage is routed via a preset level control, another record/play switch section, and then to pin 5 of IC101. Refer to the complete circuit diagram in Fig. 6-7.

204

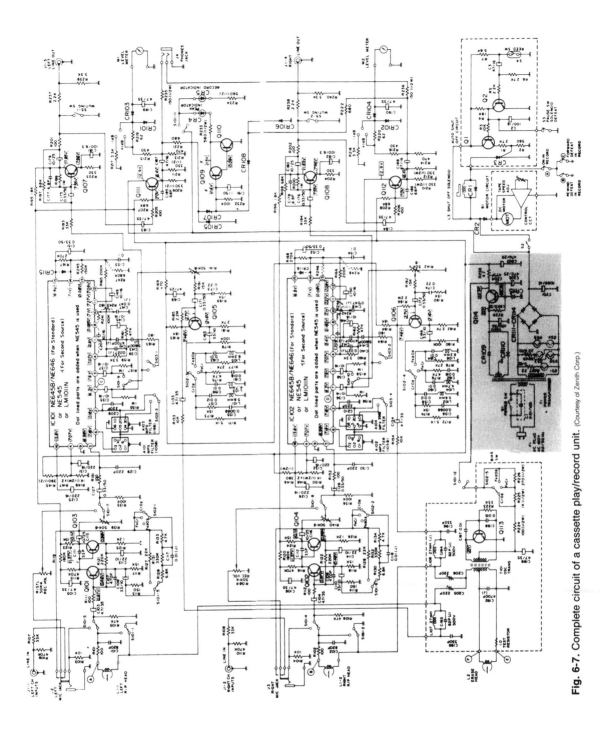

Fig. 6-7. Complete circuit of a cassette play/record unit. (Courtesy of Zenith Corp.)

IC101 provides both record and playback amplification as well as the circuitry for Dolby noise reduction functions. Between pins 5 and 6, is preamp "A". Externally between pins 6 and 2 are two low-pass filters (one is switch selected) to filter out multiplex signal components, or any signals above 15 kHz. This is done to prevent possible interference from these sources. At pin 2 is preamp "B", which connects to pin 3 and mixer circuit "C" functioning as either an adder or subtractor, depending on the operating mode. The output of the mixer "C" connects to pin 7. Then the signal is coupled to another section of the record/play switch, gain buffer Q107 and to the line output jack. A mute switch on this line will short the signal to ground to avoid unwanted signals. This switch will remove the mute action only when the play button is activated.

Switching Dolby noise reduction to ON while in the PLAY position, will result in added operating circuits which control the dynamic processing characteristics. The signal will also be routed from pin 7 to the second section of the noise reduction switch and then to a high-pass filter connected externally between pins 4 and 1. Internally, between these pins is a variable resistance "D" which has a high resistance at low signal levels. Output of this filter goes to amplifier "D" and then to mixer "C", where it is combined with the original signal processed to pin 7, and then to Q107. In addition, the signal from amplifier "E" is fed to amplifier "G" and is then rectified by a nonlinear integrator "H" to provide a dc control voltage. This dc control voltage will vary resistance "D" in such a way that when the control voltage exceeds a certain level, resistance "D" will start to fall, causing an increase in the turnover frequency of the high pass filter. This filter will then attenuate low and mid range frequencies in the signal path processed via amplifier "E". Nonlinear integrator "H" functions in a manner that will provide fast gain change action without creating unwanted signals. If transients of excessive levels should occur they will be clipped by clipper "F" resulting in a smooth response action.

The Record Mode of Operation

During the record mode, microphone signals will be processed via the R/P preamp consisting of Q101 and Q103 and is similar to that explained in the play mode of operation. External signals will be routed from the line input. Signals will then be processed via switching contacts on the microphone jack, the R/P switch, IC101 pin 5, amplifier "A", pin 6, low-pass multiplex filters, pin 2 amplifier "B" to pin 3, as during play. The signal will also be applied to mixer "C".

When noise reduction is on, while in the record mode, the signal will be taken from pin 3, through the record/play switch, then routed via the noise reduction switch and applied to the high-pass filter at pins 4 and 1. This high-pass filter will function in a manner similar to that explained for play, but now as part of a positive feedback loop (using mixer "C" as an adder) instead of the negative feedback loop used in play. This will result in record circuit characteristics which are complementary to that of playback.

The output at pin 7 will be connected to record amplifier Q105. From there the record signal will go through a bias trap which prevents the bias oscillator signal from getting back into the record amplifier or other circuits where the bias frequency could cause undesired effects. The bias oscillator serves two functions. One function is to provide a bias to the erase head, while the other is to establish an ac (RF carrier) type bias at the record/play head.

NOISE REDUCTION SYSTEMS (TAPE RECORDERS)

Noise reduction systems can be divided into two general categories of circuits, passive type filters and active type (or dynamic) filters.

Passive Filters

A typical passive filter circuit will modify frequency response and/or eliminate noise without use of any active devices (such as transistors or IC's). This type of circuit will cause signal level loss between input and output of the filter. Typical examples would include high and low filters which roll off, or reduce signal level at the high and/or low frequency ends of their response curve. A typical response curve is shown in Fig. 6-8, where the solid line represents the "wide" high end frequency response of a circuit with the high filter switch in the off position. Reduced high frequency response, shown by the dashed line, occurs when the high filter switch is turned on. While high filters are used to attenuate, or suppress, high frequencies, a low filter system can thus provide the same function at low frequencies. Although these filters are practical, in that they can suppress some noises in a circuit, they will also suppress the level of the desired signals. Out goes the noise, such as tape hiss or record surface noise, and also, out goes the high frequency content of the recorded music.

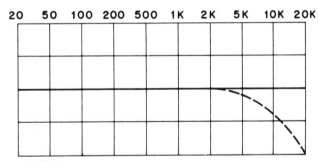

Fig. 6-8. Passive HI-CUT filter curve. (Courtesy of Zenith Corp.)

Single-Ended Noise Reduction Systems (Active)

Typical single-ended noise reduction circuits could be described as deemphasis circuits in which changes of roll-off points and degrees of attenuation vary, based on the levels and high frequency content of a signal. For example, if there is no high frequency content, the response curve will show a reduction in bandwidth. There will also be a reduction in reproduced hiss or related noise (improved signal to noise ratio).

Behavior of a typical single-ended circuit can be illustrated as shown in Fig. 6-9. A high level wide-band signal (Curve A) shows wide-band response. As the signal level is reduced, the amount of deemphasis is increased, Curves B, C, and D, respectively.

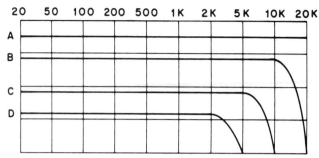

Fig. 6-9. Noise reduction response curve. (Courtesy of Zenith Corp.)

Noise reduction control in single-ended systems occurs entirely during playback. If you reproduce continuous low level, low frequency program material, this type of circuit has a tendency to mute low level high frequencies that may exist in the same program material.

It is possible to process (encode) a signal before recording, and process (decode) it again during playback, in a manner which will improve noise reduction. This can be accomplished by using a compressor/expander circuit technique and is sometimes referred to as a double-ended noise reduction circuit system.

AUTO SHUTOFF CIRCUIT (CASSETTE PLAYERS)

This circuit is used in many various cassette players to shut the machine off when end of tape is reached. A permanent magnet is mounted at one end of the tape counter shaft. On the auto-shutoff circuit board (see Fig. 6-10) is a magnetically activated reed relay mounted in a position near the magnet on the counter shaft. As the shaft rotates, the magnet will cause the contacts on the reed switch, S4, to open and close. When switch S4 is open, B+ will charge C3, via R7 and R6 to ground. When the switch is closed, C3 will be discharged through R6 and the switch contacts. The Q2 base will see

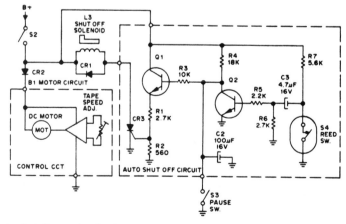

Fig. 6-10. Auto shutoff recorder circuit. (Courtesy of Zenith Corp.)

208

this action as a series of pulses of short duration which will be sufficient to turn Q2 on and keep it on. With Q2 turned on, Q1 will be cut off, preventing SCR3 from firing and triggering the shutoff solenoid.

When the end of the tape is reached, the counter shaft will stop rotating, reed switch, S4, will stop switching, Q2 base will no longer see the pulses and will cut off. This in turn will cause Q1 to turn on and current will flow in Q1. Voltage from the tap between R1 and R2 in the emitter of Q2 will be sufficient to forward bias SCR CR3. CR3 will conduct, completing the path from B+, through the solenoid and CR3, activating the solenoid. Solenoid activation causes all mechanically linked levers to go to the off position on the cassette keyboard.

8-TRACK TAPE PLAYER/RECORDER SYSTEM

The 8-track tape units use a cartridge like the one in Fig. 6-11. The tape is about 0.244 inch, which is slightly less than ¼ inch wide. We see in Fig. 6-12 the illustration for the track arrangement of a two-channel tape. As can be seen each track is about 0.022 inch wide, with a guard band of about 0.010 inch between each track. This results in the formation of 4-two-channel programs which are played back as follows.

PROGRAM	TRACKS
1	1 and 5
2	2 and 6
3	3 and 7
4	4 and 8

Therefore tracks 1, 2, 3, and 4 contain left channel information, while tracks 5, 6, 7, and 8 contain right channel information.

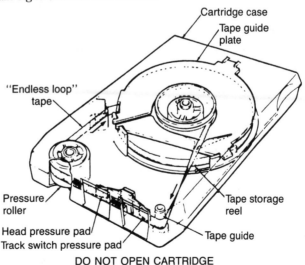

DO NOT OPEN CARTRIDGE

Fig. 6-11. Inside view of an 8-track cartridge. (Courtesy of Zenith Corp.)

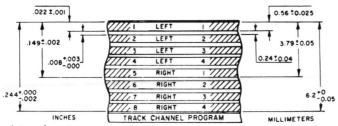

Fig. 6-12. 8-track tape format. (Courtesy of Zenith Corp.)

The 8-track cartridge units operate with the tape moving at 3 and ¾ inches per second. The tape feeds from the center of the tape reel, out to, and past, the tape guide, pressure pads, pressure roller, and back to the outside of the reel. Movement of the tape stops when the cartridge is pulled out of the tape player unit.

Let's now look at the electronic circuit operation for one type of these 8-track player units. The 8-track players can be grouped into the following systems.

- Playback only
- Play/Record-Full Feature
- Play/Record-One Button

Each unit contains the required circuitry such as a preamp circuit for each channel with built-in frequency compensation network to provide the equalization required for proper playback and/or record response.

Two channel (stereo) tape players contain a single circuit board on which the electronic components are mounted. The circuit for an 8-track player unit is shown in Fig. 6-13. The circuit contains two separate preamplifier circuits terminated into two separate outputs for connection to the audio output stages of a stereo power amplifier system.

A full feature 2 channel play/record unit (see complete circuit for this unit in Fig. 6-14) has the preamplifier circuits, VU meter circuit, and bias oscillator module on one circuit board. The bias oscillator is not adjustable and if a fault occurs, replacement of the complete module is required. A second circuit board contains the power supply components and a relay. The relay, when activated, removes the ground path from the motor and applies it to the ready indicator lamp.

Two channel play/one button record models have a non-repairable bias oscillator unit on the preamplifier circuit board. A relay which switches the ground path for the motor to the auto stop indicator lamp is also located on the main circuit board, or on a separate power supply circuit board. Two additional circuit boards, indicator panel wiring and in/out panel wiring, are also mounted within some units.

COMPACT DISC (CD) AUDIO PLAYERS

You may recall that the audio CD (compact disc) was developed by Philips, the Dutch electronics company. The laser pickup system had been developed for the video-

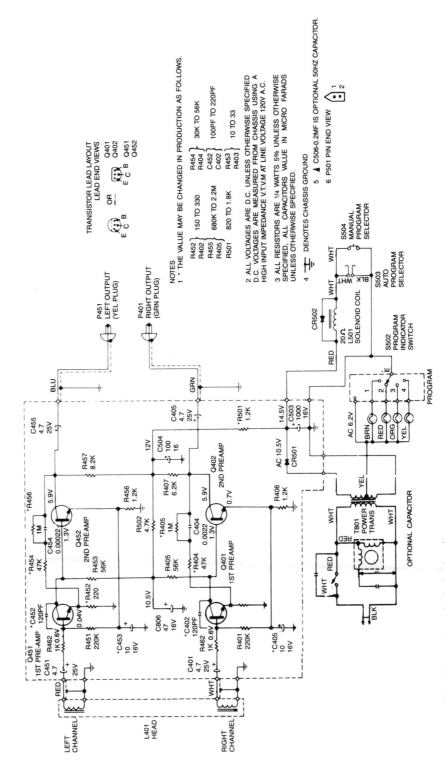

Fig. 6-13. Circuit for an 8-track playback unit. (Courtesy of Zenith Corp.)

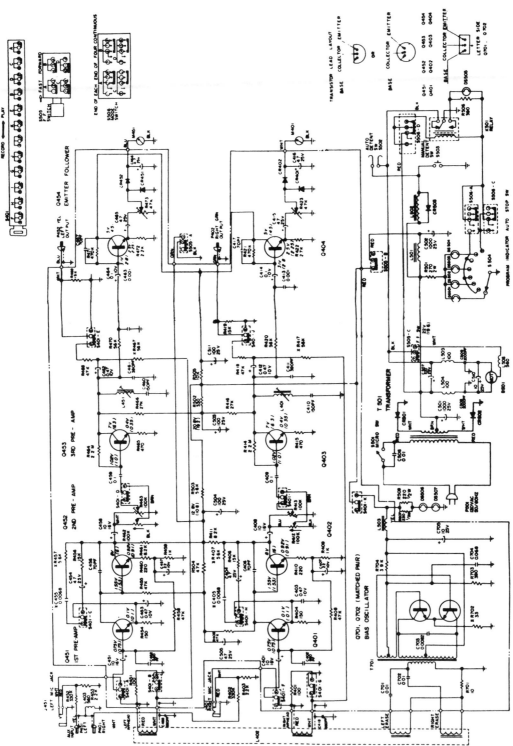

Fig. 6-14. Complete circuit diagram for an 8-track tape player/recorder unit. (Courtesy of Zenith Corp.)

disc and with some modification, adapted to audio players. The big change was to use a semiconductor laser instead of the gas version. This cut the size and cost of the CD.

The information recorded on a videodisc is analog, not digital. Around 1980, large scale IC's were possible for digital audio processing using a small number of IC's. Thus, the CD was born.

The compact disc is 120 mm (4.75 inches) in diameter, recorded on only one side, and can hold about 550 Mbytes of information. Maximum playing time is over 70 minutes, but most CD's are recorded for one hour. The information tracks are about 0.5 microns (0.005 mm) wide, spaced 1.6 microns apart. To get some perspective, a strand of hair would be close to 40 tracks wide. The actual program area is 33 mm wide and can hold up to 20,000 tracks. This calls for a very precise servo tracking system.

The disc motor for a CD player varies from a maximum of 450 rpm when the laser pickup is in the center to 240 rpm at the outside of the disc. The CD needs to see a *constant linear velocity* (CLV) during operation. Thus the CD requires that the bit stream (data) from the disc be fed into the player at a constant rate. As the pickup tracks from the inside out (unlike a conventional phonograph), it covers more linear distance each rotation because the circumference of the track becomes larger. Unless the motor is slowed down proportionally, the data rate would increase and this would upset the precise timing circuits of the CD player.

The information on the disc is encoded as a series of pits and bumps on a reflective metallized coating that is bonded to the plastic disc. Note the drawing in Fig. 6-15. A layer of clear plastic 30 microns thick is laminated onto the information side that is the

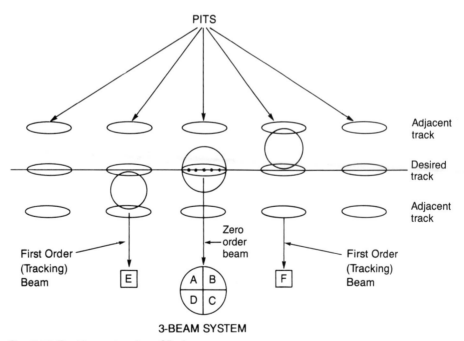

Fig. 6-15. Tracking system for a CD player.

bottom side of the disc. The pits and bumps amplitude modulate the laser light that is reflected into the pickup assembly. Tracking information and program info is derived from this data.

The signals from the disc are processed through three different circuits, to extract the FE (focus error), TE (tracking error), and EFM (music) data. Note the drawing of pickup circuits shown in Fig. 6-16. The optical pickup is usually divided into six photodetectors. A, B, C, D are the main spot detectors, while E and F are the side spot or tracking detectors. The EFM signal is equal to A + B + C + D. To generate a focus error, use FE = (A + C) minus (B + D). Note that FE will be zero whenever the terms within the parentheses are equal. This only happens when the light impinging upon the pickup forms a circular pattern, indicating it is in focus. Unevenness in the disc surface and drive system will change the distance between the lens and the information layer, making the spot oval-shaped. This produces a focus error, driving a coil that moves the objective lens on its vertical axis until correct.

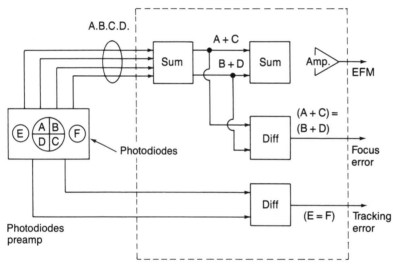

Fig. 6-16. Photodiode signal processing.

The D and F photodiodes are aimed at the tracks to either side of the current information track. They should receive equal amounts of light when the pickup is centered. Since TE = E − F, the result will be zero for these conditions. Any variation indicates mistracking, which is corrected by a coil that moves the lens laterally. When the lens has moved to its limit, the sled motor takes over momentarily, moving the entire optical unit in the direction of play. The sled takes coarse tracking and the lens fine tunes the tracking.

Another servo is required to keep the disc motor locked to the incoming data stream. The most common way is to use a PLL (phase-locked loop) to synchronize a VCO (voltage controlled oscillator) to the incoming EFM. A divider is used to generate a write-frame clock from the VCO, which is then compared to a reference clock. A phase comparator combines these two signals to produce the correction voltage for the disc motor servo.

ANALOG TO DIGITAL BITS AND EFM ENCODING

As you may know the basic concepts of changing analog to digital bits is quite simple. Just sample the analog signal level, assign a digital value to it and keep repeating. For CD's this process is repeated at a rate of 44.1 kHz, or every 22.7 microseconds. Each complete sample is represented by a 16-bit word. Left and right channels are recorded alternately.

In the real world of CDs this may not be so simple. You will find that over three-fourths of the circuitry used in the CD signal path is required to overcome hardware limitations. Also, you will probable notice that approximately two-thirds of actual information on a compact disc is there for the same reasons. Less than half of the disc material is actual program or music information.

Let's see why this is so. The first reason is that the laser beam has a finite width, in the area of 1.7 microns, which is large compared to the 0.5-micron track width, and the 1.6-micron spacing between tracks. The next reason is, to jam as much data as possible onto the disc, engineers have agreed on a NRZ (*non-return-to-zero*) data format. Refer to the drawing shown in Fig. 6-17. Just keep in mind at this time that it results in pits that the laser cannot read because they are the wrong size. This creates the need for a further level of complication—EFM.

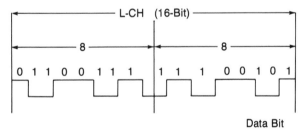

Fig. 6-17. With the NRZ (non-return-to-zero) data format, the binary value of the data stream only changes to the opposite state when a 1 is encountered.

EFM Encoding

Eight-to-fourteen modulation (EFM) ensures that all the little pits on the compact disc will be within the right size range for the laser to read. We cannot have any that are too small, because the laser will not be able to resolve them. If the pits are too big, the output of the photodetectors will veer toward dc and disturb the data clock. Note the waveform drawing shown in Fig. 6-18. We take the 16-bit words derived from the sampling process and split them into two 8-bit bytes. Now, take these and look them up in a table of 14-bit values, selected from among the 16,384 total possibilities, on the basis of what pit size they produce. The laser design makes it not want to see any digital signal with more than 10, or less than two, zeros between any two "1" values. The 14-bit model offers 267 of these.

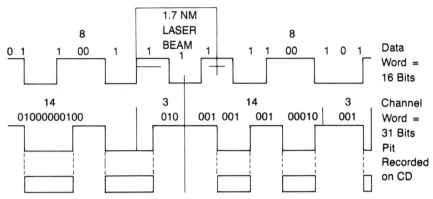

Fig. 6-18. EFM is necessary because of the NRZ format. The beam is too wide to resolve rapid transitions in a straight 16-bit system, so specially selected 14-bit patterns are substituted for each 8-bit byte. They are separated by three merging bits.

The NRZ is a digital coding scheme in which the binary value of the data stream only changes when a 1 is read. The signal does not return to zero volts after each high, the purpose being to reduce a signal's high-frequency components and thus the bandwidth requirements of the processing circuitry.

In summation, the CD recording system breaks each 16-bit word, representing one sample of the audio input, into two 8-bit bytes. Then the EFM encoder substitutes an appropriate 14-bit value for each one and sends the results off for more processing. At this point the CD has taken the original 16-bit word and derived two 14-bit patterns from it for a total of 28 bits.

This will increase the number of bits per sample from 16 to 28, a 75% increase. Thus, the actual music data is only about 57% of the EFM signal. And, of course there are other control and sync bits, too.

CD SERVO LOOP TROUBLESHOOTING

The CD player has many servo systems and troubleshooting them can be a big problem. Let's hope the following information will make servo loops easier for you to troubleshoot.

A servo is a circuit that uses some of its output as an input. This technique is similar to oscillators, op amps, and audio amplifier circuits in operation. A certain amount of the output is fed back to the input in order to keep the circuit in operation as with an oscillator or cancel distortion in an audio op amp circuit. However, servos are usually part of an electromechanical device that has a motor to control. A servo system functions to maintain the correct operating position of a mechanical assembly by comparing a position signal (feedback) to a reference signal. In the case of a CD player, it has tracking/sled, focus, and disc motor (CLV) servos. The tracking coil and sled motor are separate, but they share a good portion of the servo circuitry for their operation. In simple terms, the sled takes care of the coarse control, and tracking the fine control of the optical block.

Because of safety shutdown devices, the most common symptoms found in home CD players is a dead disc motor. Many of these units are designed so that focus must be achieved (FOK) in order for the motor to start up. However, most of the portables will run the motor up as soon as the PLAY button is pushed. Sometimes, the disc will start to spin, then shut down if TOC is not being read, probably due to RF PLL or tracking problems. One of the first checks you need to perform on a defective CD player is check the RF (eye pattern), because you are in trouble unless you can get the disc motor to run. No rotation results in no eye pattern.

You will find that some CDs have built-in diagnostics or something that allows you to defeat shutdown. Some CDs have a pin on the microprocessor that eliminates certain shutdown processes when grounded. However, the service info may not explain just what is being defeated. In most cases the TEST or ADJUST modes will keep the CD going so that you can adjust or troubleshoot the machine.

Let's go back to the CLV motor that will not run, the laser needs to be checked out. Just a few circuit jumpers or rewiring should do it for these checks. Refer to the drawing in Fig. 6-19. Many CLV motors are of the brushless, slotless type, which among other things indicates that speed control is usually affected by changing the bias on a HALL-effect device. Assuming the power to the motor is OK, you can disconnect the speed control input from the servo circuitry and substitute in a variable external power supply. Bias voltages will usually be in the zero to two volts dc range, and probably negative. Be sure and use the circuit schematic for the CD you are repairing.

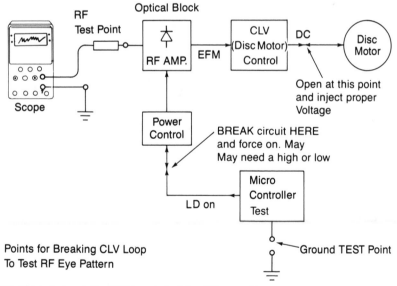

Fig. 6-19. Points to break the CLV loop to test the RF eye pattern.

You will need very precise control for the motor speed and this calls for a stable bias voltage. A power supply with precision control below 6 volts is desirable. The Hall devices require voltage and not current for operation.

On some of the portable CD players you may find the brush and commutator type of dc disc motor. These motors will have only two wires. The servo provides both power and control operation. The BSL motors only require a control voltage.

Making the motor run is just one step in the process. In order to see an eye pattern, the laser must be on and the focus servo operational. You will find that the microprocessor usually feeds some sort of enabling signal to the optical block. Even with the test mode engaged, it is often necessary to force this signal to the proper level to get the laser to stay on long enough.

With the laser turned on, and the disc motor running at approximately 300 rpm, you should obtain some type of eye pattern, if the disc is in the machine, the RF amp is working and the focus servo is working. If the focus is poor it will make the test impossible. However, there may be enough slop in the sled assembly that you can move the optical block enough to find a good focus point. And this will just have to be a moment. If the RF is good, you can assume that the laser is not the cause of the motor problem. Also, make sure that the optical block is positioned into the disc where data density is normal. In the table of contents (TOC) and between tracks you will see a very weak pattern.

You may have to perform this troubleshooting procedure a few times before you have it down pat. Make sure you understand the circuit well enough to accurately guess exactly where to break the loop and inject voltages.

RF PLL CHECKS

Let's move from servos onto PLL (*phase-locked loops*). The RF PLL is responsible for creating the *write frame clock* (WFCK), which is phase locked to the CD's data stream. By comparing WFCK to the read, or *reference frame clock* (RFCK), (derived from a crystal-controlled oscillator) the CLV circuitry generates a correction signal to control the disc motor, thus keeping it locked to the recorded data.

If speed control is good, but there are problems in the phase servo, the audio output will probably be muted. Some CDs disable muting when in the adjustment mode, however, allowing you to hear the odd sounding distortion typical of this failure. Note the simplified circuit in Fig. 6-20.

You can quickly check the PLL's ability to track a signal even if the laser is out and there is no RF (eye pattern). We will use a CD with a CX20109 RF amplifier along with a CX23035 CLV digital signal processor chip. With the unit in PLL adjustment mode, use an RF generator to feed in about a 2.16 MHz, 400 mV signal into the CX20109, pin 2 (RF1) and monitor the *phase-lock clock* (PLCK). Varying the input frequency slightly should cause PLCK to track. If not, check the EFM signal path, and also the varactor control circuitry.

Feed a signal in at the RF amplifier so that it is squared up and looks like EFM to the CLV/signal processor. For some reason the 4.3218 MHz signal does not work for this check.

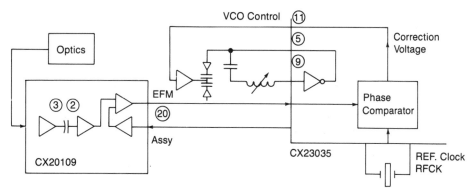

Fig. 6-20. This circuit shows how to check the RF PLL. With the unit in PLL adjustment mode, use an RF generator to feed a 2.16 MHz, 400 mV signal into CX20109, pin 2 (RF1) and monitor the phase-lock clock (PLCK). Varying the input frequency slightly should cause the PLCK to track.

If the VCO section of the loop is inoperative, you can test the inverter inside the CX23035 by toggling pin 9 with a logic high or low, while you watch for inversion at pin 8. If you find no inversion present this usually indicates a bad chip.

Tracking Down Electronic or Mechanical Troubles

Probably the most common symptom of CD tracking problems is skipping or mistracking. It sounds very much like the problems that will occur on the old time analog phonograph records of yesteryear.

You, the technician will have to determine if the problem is electronic or mechanical in nature. A complete check and touch-up of all tracking-related adjustments is usually all that is required. The most critical ones, you will probably find, are tracking balance (E-F balance) and RF PLL. It is important that the output of the tracking detector amplifiers are symmetrical around the 0 volt dc, (positive and negative peaks are equidistant from the 0 volt baseline), so that the output is truly zero when there is no tracking error.

Proper adjustment of the RF PLL assures that the disc motor cooperates with the optical block, and is especially important during dropouts caused by defects on the disc. If the free-run frequency of the PLL is too far from 4.3218 MHz, it will take longer to recover after a dropout.

Is It the Sled or Optical Block?

Many times all of the adjustments look good, the eye pattern is good, but the player still skips around intermittently. Now the question, is it the optical block or the sled drive mechanism? Use the scope to test for tracking errors at test point (TE) set for a time base sweep of 5 ms. You should see what looks like random noise. However, should mechanical binding occur in the sled drive, the gradual buildup of a fundamental frequency in the range of 250 Hz will be seen on the scope. To see how this looks on the scope, use an operating CD player and jam the sled during play as you monitor at test point TE.

Another way is to check the actual drive going to the sled motor. If it uses pulse-width modulation (PWM) drive, it is very easy to detect mechanical binding. As an example, the sled drive pulsewidth on a Sony D7 typically varies between 2 μs and 7 μs. When there is too much friction in the drive system, the pulse width slowly increases, then jumps back to zero right after a skip. This occurs as the drive circuitry tries to pump even more power into the motor.

Either the sled assembly breaks loose and jumps forward, or the objective lens reaches the end of its range and jumps backwards. In either case, for a moment there is zero tracking error, and thus no correction. This is a very good indication of a mechanical problem.

SONY CD PLAYER

The front panel operating controls for the SONY model CDP-C70 CD player is shown in Fig. 6-21.

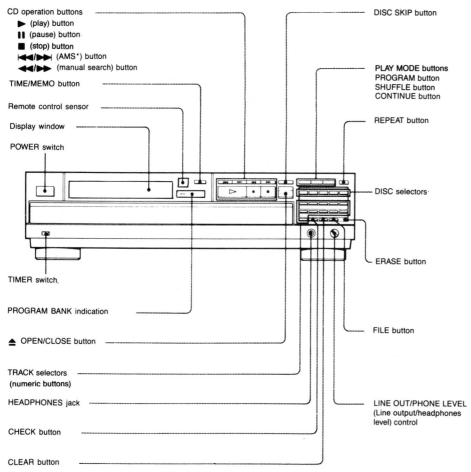

Fig. 6-21. Front panel controls of SONY CDP-C70 CD player. (Courtesy Sony Corp. of America)

Sony CD Electrical Adjustments

NOTE: Perform adjustments in the order given.

1. Use YEDS-18 disc unless otherwise indicated.
2. Use an oscilloscope with more than 10 Meg. Ohm input impedance.

RF PLL Frequency Adjustment/Lock Frequency Check Procedure:

1. Connect test point TP (ASY) to ground with the clip lead.
2. Turn the POWER switch ON.
3. Connect the frequency counter to test point TP (PCK). Note the counter connections shown in Fig. 6-22.
4. Adjust RV231 so that the reading on the frequency counter is 4.3218 MHz ± 30 kHz. (RF PLL frequency adjustment).
5. Remove the clip lead connecting TP (ASY) to ground.
6. Insert the disc (YEDS-18) in and press the button.
7. Confirm that the reading on the frequency counter is 4.3218 MHz.

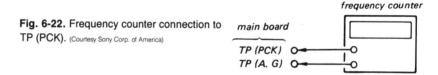

Fig. 6-22. Frequency counter connection to TP (PCK). (Courtesy Sony Corp. of America)

E-F Balance Adjustment Procedure

This adjustment should be made when replacing the TOP (T-type Optical Pickup).

1. Connect test point TP (ADJ) and test point TP (TES) to ground with the clip lead.
2. Connect the oscilloscope to test point TP (TEO). Note Fig. 6-23 for scope connections.
3. Turn the POWER switch ON.
4. Insert the CD disc (YEDS-18) in the press the button.
5. Adjust RV101 so that the traverse scope waveform is symmetrical above and below as shown in Fig. 6-24.
6. After adjustment, remove the jumper clip lead that was connected at TP (TES) to ground.

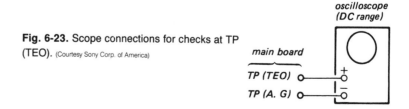

Fig. 6-23. Scope connections for checks at TP (TEO). (Courtesy Sony Corp. of America)

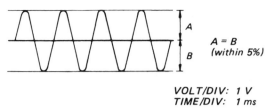

$A = B$
(within 5%)

VOLT/DIV: 1 V
TIME/DIV: 1 ms

Fig. 6-24. Correct symmetrical waveform. (Courtesy Sony Corp. of America)

Focus Bias Adjustment Procedure

This adjustment should be performed when replacing TOP (T-type Optical Pickup).

1. Connect the oscilloscope to test point TP (RF) and test point TP (RF.G). Refer to Fig. 6-25.
2. Turn the POWER switch ON.
3. Insert the disc (YEDS-18) in and press ▷ button.
4. Adjust RV104 for an optimum waveform eye pattern or so the peak is at a maximum. Optimum eye pattern means that the shape "◇" can be clearly distinguished at the center of the waveform. Note the scope pattern shown in Fig. 6-26.

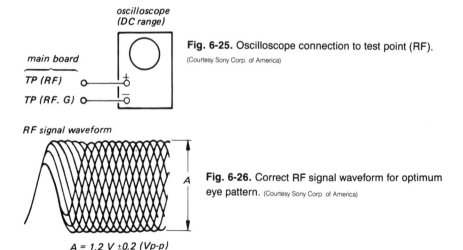

oscilloscope
(DC range)

main board

TP (RF)

TP (RF. G)

Fig. 6-25. Oscilloscope connection to test point (RF).
(Courtesy Sony Corp. of America)

RF signal waveform

A

Fig. 6-26. Correct RF signal waveform for optimum eye pattern. (Courtesy Sony Corp. of America)

$A = 1.2 \ V \pm 0.2 \ (Vp\text{-}p)$

Focus/Tracking Gain Adjustment

A frequency response analyzer is necessary in order to perform this adjustment correctly. However, the gain has a margin of error, so even if it's off slightly, there is no problem. You may want to skip this adjustment.

Focus/tracking gain determines the pickup follow-up (vertical and horizontal) relative to mechanical noise and mechanical shock when the 2-axis device operate. However, as these reciprocate, the adjustment is at the point where both are satisfied.

- When gain is raised, the noise while the 2-axis device operates increases.
- When gain is lowered, mechanical shock and skipping occurs more easily.
- When the gain adjustment is off, the symptoms shown in the Table 6-1 will appear more often.

Table 6-1. Symptoms Chart. (Courtesy Sony Corp. of America)

Gain Symptoms	Focus	Tracking
• The time until music starts becomes longer for STOP → ▷PLAY or automatic selection (◄◄ ►►I buttons pressed.) (Normally takes about 1 second.)	low	low or high
• Music does not start and disc continues to rotate for STOP → ▷PLAY or automatic selection. (I◄◄ ►►I buttons pressed.)	—	low
• Disc table opens shortly after STOP → ▷PLAY.	low or high	—
• Sound is interrupted during PLAY. Or time counter display stops progressing.	—	low
• More noise during 2-axis device operation.	high	high

Primary Adjustment Procedure

NOTE: Since exact adjustment cannot be performed, remember the positions of the controls before performing the adjustment. If the positions after the primary adjustment are only a little different, return the controls to their original position.

1. Keep the CD player horizontal. (If the CD player is not horizontal, this adjustment cannot be performed due to the gravity against the 2-axis device).
2. Insert the CD disc (YEDS-18: Fifth Selection) and press the PLAY button.
3. Connect the oscilloscope the main amplifier board at TP (FEO-FEI). Note the scope setup shown in Fig. 6-27.
4. Adjust RV103 so that the waveform is as shown in Fig. 6-28. The focus gain adjustment.

Some incorrect scope waveform examples are shown in Fig. 6-29. The dc levels change more than they normally should.

Procedure:

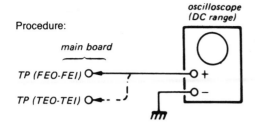

oscilloscope
(DC range)

main board

TP (FEO-FEI)

TP (TEO-TEI)

Fig. 6-27. Scope connections to main amplifier board. (Courtesy Sony Corp. of America)

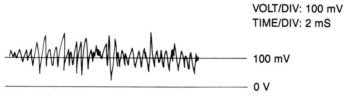

VOLT/DIV: 100 mV
TIME/DIV: 2 mS

100 mV

0 V

Fig. 6-28. Correct waveform for focus gain adjustment. (Courtesy Sony Corp. of America)

low-focus gain

VOLT/DIV: 100 mV
TIME/DIV: 2 mS

250 mV

0 V

high-focus gain

VOLT/DIV: 100 mV
TIME/DIV: 2 mS

100 mV
75 mV
0 V

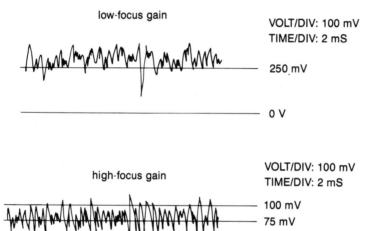

Fig. 6-29. Waveform for low- and high-focus gain. (Courtesy Sony Corp. of America)

5. Connect oscilloscope to main board TP (TEO-TEI) as shown in Fig. 6-32 for its location.

6. Adjust RV102 so that the waveform is as shown in Fig. 6-30. An incorrect scope waveform is shown in Fig. 6-31. Note that the fundamental wave appears.

VOLT/DIV: 1 V
TIME/DIV: 2 mS

0 V

Fig. 6-30. Correct waveform. (Courtesy Sony Corp. of America)

224

VOLT/DIV: 1 V
TIME/DIV: 2 mS

0 V

Fig. 6-31. Incorrect scope waveform. (Courtesy Sony Corp. of America)

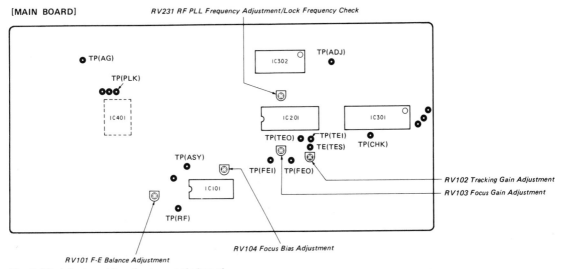

[MAIN BOARD]

RV231 RF PLL Frequency Adjustment/Lock Frequency Check

TP(AG)

TP(ADJ)

IC302

TP(PLK)

IC401

IC201

IC301

TP(TEO)   TP(TEI)
TE(TES)   TP(CHK)

TP(ASY)

TP(FEI)   TP(FEO)

IC101

RV102 Tracking Gain Adjustment
RV103 Focus Gain Adjustment

TP(RF)

RV104 Focus Bias Adjustment

RV101 F-E Balance Adjustment

Fig. 6-32. Adjustment locations on main board. (Courtesy Sony Corp. of America)

7

Video Cassette
Recorder (VCR) Circuits

In this chapter we will look at selected circuit operations found in the Sony Betamax VCR system and the RCA (VHS) Home Video Recorder system. Plus, Sony Super Beta Circuits and 8 mm tape format.

RCA VCR CIRCUIT OPERATIONS

We will now look at video recorder circuits found in some model VBT 200 RCA home video recorder systems.

The Power Supply Circuit

The video cassette recorder power supply shown in Fig. 7-1 uses a power transformer (T6301) and two bridge-rectifier circuits to develop sources of unregulated 12 volts dc and unregulated 18 volts dc. A third unfiltered supply, using diodes D107 and D108, produces the "power-off detector" supply. The voltage is fed to the regulator and transport board logic system to sense a power failure and operate the stop solenoid so that the machine is not left in an operating mode in the event that ac power is lost. A second power transformer (T6302) supplies 16 volts ac and 3 volts ac to operate the digital timer.

Ac power for the main power transformer (T6301) and the timer power transformer is supplied via the ac line with a 1.6 amp fuse. Because of the low power demand of the clock/timer, a separate 100 mA fuse is used to protect the timer power transformer. Other fuses protect the rectifier/filter system and the unregulated 18 volt supply.

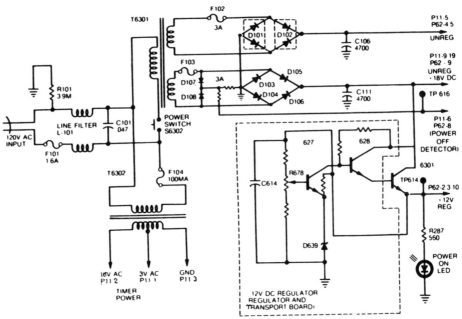

Fig. 7-1. Power supply circuit with +12 volt regulator. (Courtesy of RCA)

The unregulated 12 volt supply output is fed to the regulator and transport board where it encounters some switching. This voltage is also fed to the D-D motor board where it provides input power to operate the three phase inverter that then drives the D-D cylinder motor.

The unregulated 18 volt supply is also fed to the regulator and transport board where it provides power to operate much of the logic circuitry contained on this board as well as the stop solenoid. Also derived from the unregulated 18 volt supply is regulated 12 volt dc which is generated by a series-regulator circuit that utilizes low-level drive circuitry on the R & T board, and a power transistor (TR6301) which is chassis mounted. The regulated 12 volt source appears at the emitter of transistor TR6301 and can be measured at test point TP614. Also, note that the "power-on" indicator (an LED) is powered from the regulated 12 volt source. Thus, from the servicing standpoint, if stop-solenoid action is heard when the VCR power is turned "on" and "off", but the "power-on" indicator does not come "on", this would point to a problem in the regulator circuit. This is because the 18 volt supply powers the stop solenoid and supplies input voltage to the regulator which then drives the "power-on" LED. Most of the circuitry in this VCR is powered from the regulated 12 volts power source.

In summation, the power supply circuit provides three voltages to the R & T boards. These voltages are +12 volts unregulated, "power-off" indicator voltage, and +18 volts unregulated. These are all directed to the R & T board. A driver circuit on this board supplies bias to the 12 volt regulator transistor (TR6301) which is chassis mounted. Also, associated with plug P61, pin 3 is the regulated output of 12 volts from this board which is fed to the luminance subprocess board via plug P39-5.

Power Supply Service Tips

Because the power supply is the heart of any electronic system, if you have a VCR that is completely inoperative or some functions do not work at all, then check out all power sources for the correct voltage.

If some dc voltages are low or fuses are blown then look for leaky filter capacitors or shorted diodes in the bridge rectifier circuits. Also, be on the alert for shorts or circuit loading from other VCR system circuits. If you suspect shorts in other sections of the VCR, then disconnect each supply voltage lead, one at a time, until the excessive load is located.

Should the +12 volt regulated power source not be correct, measure the +18 volt dc and see if it is correct. If the +18 volt power source is correct then look for component trouble in the 12 volt regulator board.

Chroma Record Circuits (RCA Home VCR)

The chroma record circuits are separated from the luminance processing as can be seen in Fig. 7-2. Video signals enters the luminance board via P31-1 where it can be

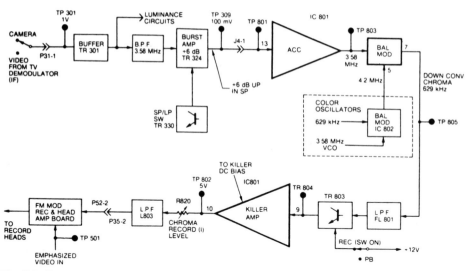

Fig. 7-2. Simplified chroma record block diagram. (Courtesy of RCA)

observed as 1 volt signal at TP301. This signal passes via the isolating buffer TR301 and is routed to the luminance and chrominance processing circuitry. Note that the video signal output of buffer TR301 is applied to the 3.58 MHz bandpass filter T308 which separates the chroma component from the luminance signal. The chroma signal then is applied to a stage called the "burst amplifier" which also serves as a chroma amplifier. The designation burst amplifier results from the fact that when the unit is operating in the 2-hour "record mode", the burst receives an additional 6 dB of amplification to accomplish the desired 6-dB burst enhancement function which is part of the VHS standard. In the 4-hour mode of operation, the additional burst boost is not used, and

228

transistor TR324 only serves as a chroma amplifier stage. Output from the burst amplifier is viewable at test point TP309. This signal is then routed to the chroma board via interconnecting cables.

Once the signal is on the chroma board, it enters chroma processing IC801 at pin 13. The scope can be used to view this signal at test point TP801. Once inside the IC, the chroma signal passes through ACC (automatic chroma control) circuitry which removes any level variations, in the same way as the ACC circuitry does in a color TV receiver. ACC regulated chroma is applied to a balanced modulator which is part of the same IC. The balanced modulator is pin 3 of the IC, and the signal can be seen with the scope at test point TP803. Also entering the balanced modulator via pin 5 is a 4.2 MHz CW signal generated by beating a 3.58 MHz VXO output CW with 629 kHz rotary CW in a second balanced modulator. The 629 kHz down-converted chroma signal (scope at test point TP805) emerges from the chip at pin 7.

A low-pass filter (C818 and FL801) rejects everything except the 629 kHz chroma signal. Output from this filter (selectable by an electronic switch TR803 which is "on" during "record") is routed back to IC801 via pin 9 and can be scoped at TP804. The signal entering the chip passes through a killer-amplifier stage which is turned "on" and "off" by the color killer circuit. When the VCR is recording a color program, the killer amplifier stage is active and passes a signal via pin 10. This down-converted 629 kHz signal can be scoped at TP802. After passing through the chroma-record level control (R820) and a low level filter (L803 and C807), the down converted 629 kHz chroma is fed back to the luminance board so it can be transferred to the FM-modulator record and head amplifier board via plug P35-2. Once on the FM-modulator board, the chroma signal is mixed with luminance in the record amplifier stages.

Service Tips

These record/play chroma circuits can be best serviced by using the oscilloscope to trace the signal throughout the various stages and IC's.

Record/Play Circuits (RCA Model VBT-200 VCR)

These record/play circuits are used in RCA's model VBT-200 video cassette recorders (VCR).

As we note in Fig. 7-3 the RF signal (head-amplifier output) is fed to the luminance board via plug P34-2. At this point, the RF signal is directed to the luminance and chroma circuits. Following the chroma signal path, the RF signal drives amplifier transistor TR327 whose output is applied to the ACC amplifier on the chroma board (via low-pass filter FL 308 and electronic switch transistor TR326). You may wish to check for this signal at test point TP309, with the scope.

Once the signal is on the chroma board, it can be scoped at test point TP801 as the ACC-amplifier input signal to IC801 pin 13. The ACC-amplifier system of IC801 acts to minimize level changes in the chroma signal that might be caused by different video tape characteristics and/or physical defects in the tape. Output from the ACC amplifier (TP803) is applied to a balanced modulator which is part of the same chip via pin 3. Also

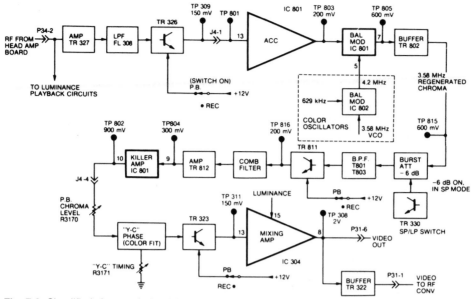

Fig. 7-3. Simplified chroma playback block diagram. (Courtesy of RCA)

entering the balanced modulator via pin 5 is a 4.2 MHz CW signal. In playback, the balanced modulator output is the difference between the 629 kHz down-converted chroma and the 4.2 MHz CW signal which is the 3.58 MHz up-converted chroma. Balanced-modulator output is via IC801 pin 7 and this signal can be seen with the oscilloscope at TP805. In playback, the 3.58 MHz regenerated chroma signal drives the buffer transistor TR802 whose output signal can be observed at TP815.

Up-converted chroma is then fed to TR807, a 2 hour/4 hour switchable 6-dB burst-attenuation stage. In the 2-hour mode, burst is attenuated by 6 dB because it was originally boosted by 6 dB during "record". In the 4-hour mode, this stage merely serves as a chroma amplifier. After passing through a 3.58 MHz bandpass filter (T801 through T803), filtered chroma (TR807 signal) drives a playback comb-filter circuit via an electronic switch transistor TR811 which is turned "on" during playback. This switch output can be scoped at test point TP816 when troubleshooting this circuit stage.

The comb filter is a rather sophisticated circuit that utilizes a "1-H delay line" in conjunction with the chroma rotary-phase recording technique to minimize chroma cross talk in the color video signal. The chroma rotary-phase system is designed to assure that the chroma cross-talk component is always out of phase with the desired signal. Because of the line-to-line redundancy of the color information in the picture, it is possible, by use of a 1-H delay line, to add the chroma signal from the chroma processing circuit along with the chroma from the previous line (1-H delayed) to obtain twice the amplitude of the chroma and cancel the cross talk. Obviously, this is a highly simplified explanation of the chroma rotary-phase and the comb-filter circuits. However, for most troubleshooting procedures, this explanation of the circuit operation should suffice.

Chroma from the comb filter drives the amplifier transistor TR812 whose output signal (TP804) is fed to pin 9, of IC801, where it passes through a killer-amplifier stage that enables or defeats the chroma circuits with the presence or absence of a color signal. Output from the killer amplifier emerges from the chip at pin 10 and is then fed to the luminance board via jumper J4. This signal can be viewed with the scope at test point TP802. Once the chroma signal is back on the luminance board, it passes through the playback level control (R3170) which allows adjustment of the playback chroma amplitude.

The chroma signal is then fed to the "Y-C" phasing circuit. Basically, the "Y-C" phasing circuit consists of phase-shift components in conjunction with the "Y-C" timing control R3171. The purpose of this circuitry is to allow adjustment of luminance/chrominance timing, thus color fit. This network output passes through a playback electronic switch (transistor TR323) whose output is matrixed with the luminance signal in the mixing amplifier of IC304. The chroma input signal can be seen with the scope at TP311. Once the chroma signal is mixed with luminance, it follows the previously described path to the RF modulator or to the external video-output jack.

ZENITH (SONY) BETAMAX VCR SYSTEM

In this section we will be looking at some selected circuits from the Zenith JR 9000 VCR machine. Let's first look at the block diagrams of the overall system operation.

Video Recording Mode

The signal is fed to the tuner and i-f board and detected for video and audio signals. The luminance signal routes the video via the Y-FM modulator block. A developed AGC signal automatically regulates the level of the *input* signal to the modulator.

The video signal is then fed through a 3.58 MHz trap circuit and now becomes the luminance signal. Luminance is then converted to an FM signal by the FM modulator and then fed to the two recording heads.

Referring to Fig. 7-4, we see the video signal is then fed on for chroma processing.

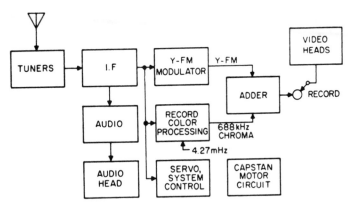

Fig. 7-4. Block diagram of record mode. (Courtesy of Zenith Corp.)

231

The chroma signal system is a "color under" system in which the chroma position of the NTSC video signal is down-converted to 688 kHz from 3.58 MHz. This is achieved by heterodyning a 4.27 MHz signal with the 3.58 MHz and using the difference signal 688 kHz. This signal is then processed to develop the automatic color control (ACC) and the automatic color killer (ACK) voltages and then coupled to the video head. These record signals are shown in Fig. 7-5.

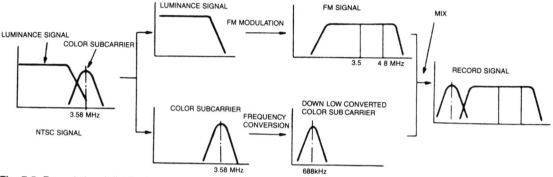

Fig. 7-5. Record signal distribution. (Courtesy of Zenith Corp.)

Both the FM luminance and the down-converted 688 kHz color signals are combined in the two video heads, which produce the recorded video signal on the tape.

The video signal input to the control circuit on the RS-L board activates the drum servo and pulse generator integrated circuits.

The servo system controls the speed of the head drum only and the pulse system produces switching signals used in switching between the two heads and in switching the 4.2 MHz carrier. The vertical signal which has been separated from the composite video is passed through an integrator network to extract only the vertical sync pulse. The output toggles a flip-flop which divides and shapes the signal into a 30 Hz square wave. This "control" signal is recorded onto the tape. This same signal is processed to control the drum brake; and thus the speed of the heads.

Video Playback Mode

On playback, the video chroma and control signals are taken from the video heads, then processed and routed out through an RF modulator with an RF output on either channel 3 or 4. Refer to Fig. 7-6 for a simplified block diagram of the playback mode.

The video signal (Y-FM) is picked up by the video heads, sent to the YC-L board where it is amplified, deemphasized, limited, demodulated, and sent to the RF modulator which gives an RF output on either channel 3 or 4.

The chroma signal pickup from the video head is converted back to 3.58 MHz. Any cross talk or jitter is attenuated and ACC and ACK voltages are developed. The chroma is then added to the luminance signal, coupled to the RF modulator, and then out of the VCR to the TV monitor.

232

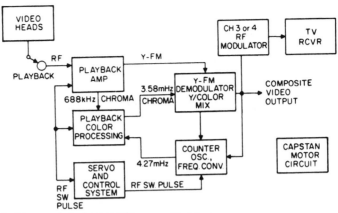

Fig. 7-6. Block diagram of playback mode. (Courtesy of Zenith Corp.)

The control track on the tape is picked up by the control head, amplified and sent to a delay circuit and onto the brake circuit. The speed of the head drum is therefore servo controlled by comparison against the playback control signal.

In early model VTRs, the rotational speed of the head drum and the forward speed of the tape past the drum was selected so that between each of the slant video tracks, recorded on the tape, is an unrecorded area. This blank area, called the "guard band" has been necessary in order to reduce cross talk between adjacent tracks during playback. The concept is the same as with the guard band between the tracks on a longitudinal audio recorder.

In the Betatape system, the guard band between adjacent tracks has been reduced to zero. Two comparatively new techniques are utilized to reduce cross talk. One principle of this technique is based upon the inherent loss of high frequency playback response which occurs when azimuth misalignment exists between the record and playback head gaps. This phenomenon is well known to most technicians who have made head azimuth adjustments in audio recorders. The angle of both the record and playback head gaps to the recorded track must be the same or high frequency response is lost.

Since in any two-head, helical scan VCR, the adjacent track on each side of the track being played back, at any given instance, is always a track recorded by the head not in contact with the tape. Thus, cross talk can be attenuated at the relative high luminance FM frequencies by deliberately introducing an azimuth displacement between the two head gaps.

APC Circuit Operation

The APC (automatic phase control) functions only in the playback mode, but not in the record mode.

A block diagram of the APC circuit is shown in Fig. 7-7. It consists of a phase detector, reactance circuit, 3.57 MHz (3.58 MHz − ¼fH) crystal controlled oscillator, burst gate circuit which performs phase detection only for the burst signal, and a 3.58 MHz crystal controlled oscillator whose output is used as a phase reference. The

233

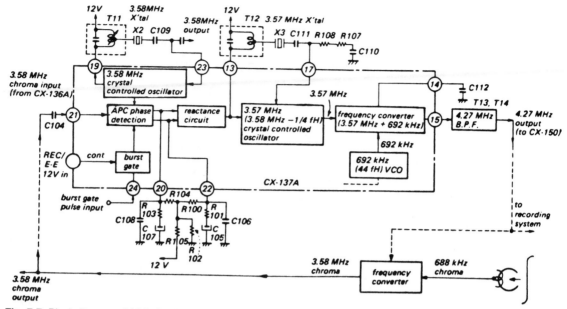

Fig. 7-7. Block diagram of APC circuit. (Courtesy of Zenith Corp.)

reactance circuit is a variable capacitance type and is connected to the 3.57 MHz crystal controlled oscillator to be part of its oscillating time constant. The reactance circuit and the crystal controlled oscillator are interconnected so that the 3.57 MHz oscillator is a voltage controlled crystal oscillator (VCXO). In the record mode the 3.57 MHz oscillator is a stable reference oscillator. In playback the frequency of the VCXO is controlled by the output of the phase detector. The VCXO output is supplied to the frequency converter where it beats against the output of the 692 kHz (44 fH) VCO to produce 4.27 MHz. This 4.27 MHz output is fed to a bandpass filter and supplied to the chroma frequency converters of the record and playback system. The playback 3.58 MHz chroma signal is then fed to the APCV phase detector. The reference signal is supplied by a 3.58 MHz crystal controlled oscillator. The output signal of this oscillator and the burst portion of the two (record and playback) chroma signals are phase compared.

The detector output is filtered in RC filters and fed to the VCXO to control the 3.57 MHz oscillator frequency. Since phase comparison is made only with the burst signal, the burst gate controls the phase detector so that the output of the APC phase detector is supplied only during the burst period. In the record mode, the output of the APC phase detector is blocked by applying the REC/E-to-E + 12 volts dc to the burst gate as a control voltage in order to stop the operation of the burst gate. This allows the 3.57 MHz VCXO to function as a fixed frequency oscillator in the record mode. In playback, the 688 kHz Chroma signal reproduced by the video heads is at a frequency which is 43.75 times the horizontal sync signal and contains phase instability caused by mechanical jitter. The AFC loop slightly over-corrects this phase instability since the horizontal sync frequency cannot be multiplied by 43.75. The APC loop detects the small phase variations which result from this over correction and develops a dc control voltage which controls the instantaneous frequency (phase) of the 3.75 MHz (3.58 MHz − ¼fH) VCXO to compensate.

CX137A IC APC Circuit Checks

For proper voltage at the chip check for +12 volts dc at pin 2 of the chip. Also, check for +8 to +8.5 volts dc at pins 20 and 22 of the IC. If voltage is not correct check components at pins 20 and 22. If components are good then suspect a faulty chip.

For APC loop operation make sure the scope waveforms at pins 4, 5, and 6 are like those shown in the Fig. 7-8 block diagram.

The Chroma Record System

The chroma record system functions as a record ACC (automatic chroma level control), as a color killer, and performs the frequency countdown conversion to 688 kHz. The block diagram in Fig. 7-9 illustrates chroma record operation. The video signal is applied to a 3.58 MHz bandpass filter where the chroma signal is separated.

The filter output is supplied to the ACC gain-controlled amplifier. This circuit is similar to the ACC control range found in a TV receiver. This amplifier controls the gain to maintain the burst signal at a constant level.

The ACC output is fed to a frequency converter and burst gate circuit. Also, applied to this circuit is a delayed horizontal sync input. It opens the burst gate at the proper time to allow only the burst through. This burst signal is then coupled to the burst transformer and then to a crystal ringing filter circuit which converts the burst signal to a CW signal. The mean amplitude of this CW signal is proportional to the amplitude of the burst signal. The filtered CW signal is then sent to an amplifier which amplifies it to a level high enough to drive the following ACC detector circuit.

The ACC peak detector detects the CW signal and compares it with the reference voltage to obtain an error voltage. The ACC output level is adjusted by changing the reference voltage. The output voltage is supplied to the ACC gain-controlled amplifier and to the color killer stage via the dc amplifier.

The color killer level is automatically set by the setting of the ACC output level, so that the killer actuates when the ACC level drops 1 dB below the reference. The color killer output is applied to the frequency converter. When no color signal is present, the killer cuts off the frequency converter. The color killer output voltage is +4 volts in the chroma mode, and zero in the monochrome mode.

The chroma signal is routed from the ACC control amplifier to the frequency converter. In the frequency converter, which is a balanced modulator, the chroma signal is heterodyned against a 4.27 MHz CW signal supplied to the frequency converter. This provides the sum and difference components of the two input frequencies. These frequencies are about 7.9 MHz and 688 kHz. The output of the frequency converter then goes to a low-pass filter which rejects the higher frequency allowing only the 688 kHz signal to pass through.

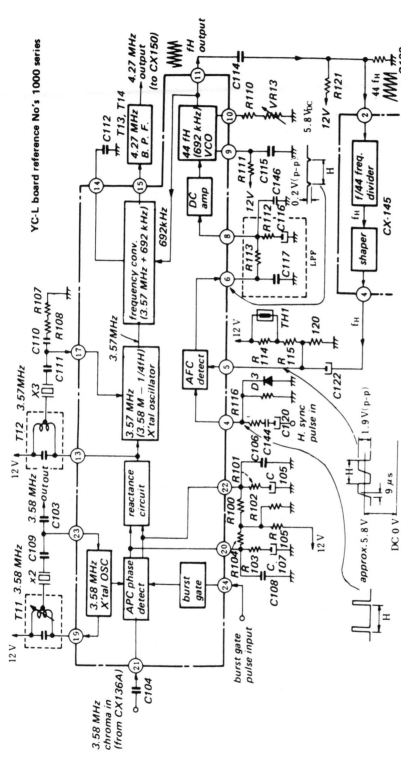

Fig. 7-8. Block diagram of CX137A chip. (Courtesy of Zenith Corp.)

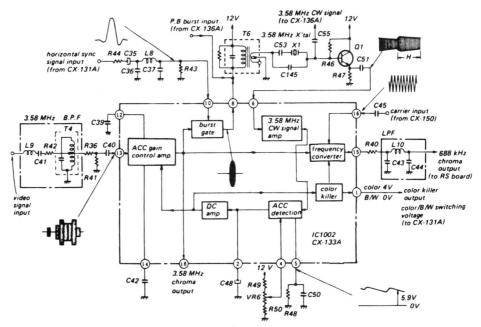

Fig. 7-9. Record chroma signal processing circuit. (Courtesy of Zenith Corp.)

Chroma Signal System Circuit

After separation from the Luminance, the 3.58 MHz Chroma signal is applied to a frequency converter as was previously stated. Here it is heterodyned with 4.267919 MHz signal. The resultant 688.3774 kHz signal is recorded on the tape. The 4.267919 CW signal is derived in a manner such that it is always phase locked to the horizontal sync in the incoming video signal. The output signal from a 3.575611 MHz crystal oscillator is beat against the output signal from a 692.307 kHz voltage controlled oscillator, and the VCO is kept phase locked to horizontal sync. When the two signals heterodyne in the frequency converter, the resultant sum frequency is the 4.267919 CW signal. The 692.307 kHz VCO is kept phase locked to the incoming "H" sync by dividing down its output frequency to 15.734 kHz in a divided by 44 countdown circuit.

The 15.734 kHz output signal from the countdown circuit is phase-compared against incoming separated H sync in a phase comparator which yields a dc output voltage. The dc output from the phase comparator is applied to the 692.307 kHz VCO as the control voltage. As shown in Fig. 7-10, the 692.307 kHz output frequency is exactly 44 fH. When they heterodyne, the resultant 4.267919 MHz signal is equal to 3.579545 MHz + (44 − ¼) fH. This means that the down-converted chroma signal has the relationship (44 − ¼) fH locked to incoming horizontal sync. As the result of this relationship, the 688 kHz down-converted chroma signal recorded on the tape retains the same relationship to H sync as did the original 3.58 MHz chroma signal. The carrier phase inverter circuit is driven by a flip-flop output and RF switching pulse. The flip-flop

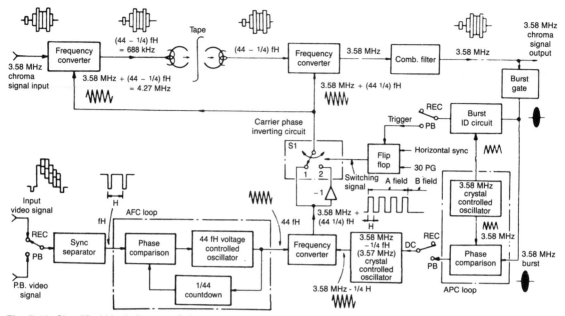

Fig. 7-10. Simplified block diagram of chroma system. <small>(Courtesy of Zenith Corp.)</small>

is triggered by the horizontal pulse. The 30 Hz switching pulse is supplied by the RG board. During "A" head field period, the flip-flop is toggled back and forth line-by-line. Then the 30 Hz pulse switches to "B" head and no phase reversals occur because the horizontal pulse cannot change the flip-flops mode.

CX133A Chroma Processing IC

The type CX133A IC has been specially developed for use in the chroma signal processing circuit of VCRs. This IC encompasses the following seven functional circuits.

- Input for 3.58 MHz chroma ACC gain control amplifier.
- Burst signal gate.
- 3.58 MHz CW (continuous wave) amplifier.
- Detector for ACC drive.
- Dc amplifier
- Color killer signal generator circuit.
- 3.58 MHz frequency converter into 688 kHz chroma signal.

Troubleshooting Tips

Since these seven circuits are working together, the individual circuits cannot be checked independently. Use the block diagram in Fig. 7-11, which gives all of the correct oscilloscope input and output waveforms along with the correct dc voltages that should be found on pins of this CX133A chip.

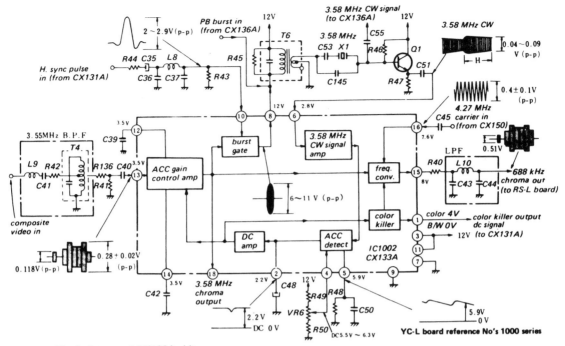

Fig. 7-11. Block diagram of CX133A chip. (Courtesy of Zenith Corp.)

Chroma Playback "Chip"

In the chroma playback mode the RF signal is fed from the RF playback amplifier on the RS-L board. This signal contains both the chroma and Y-FM signals. This signal is routed through a low-pass filter which rejects the Y-FM signal, extracting only the 688 kHz chroma signal. Refer to Fig. 7-12 for the CX136A chroma playback chip block diagram.

The 688 kHz signal is applied to an ACC gain controlled amplifier. The dc voltage, obtained by detecting the playback burst signal, is applied to the gain controlled amplifier. A feedback loop thus formed adjusts the gain of the ACC amplifier so that the burst signal level remains constant. Since separate capacity stage circuits for the ACC detection output are provided for each channel, the ACC loops are independent for each of the two heads and no difference in chroma output level is distinguishable even if a large level difference exists between the two head outputs.

The ACC output signal, 688 kHz, is applied to a frequency converter where it is heterodyned and the difference signal of 3.58 MHz is produced. The phase of the playback chroma signal is inverted at a 1H rate when the "A" head is playing back. This is restored to continuous phase in the frequency conversion process.

The signal now passes through the 3.58 MHz filter and then to an emitter follower circuit and is coupled to the comb filter, where the cross talk component is removed. The comb filter output is applied to an output amplifier. From the amplifier the signal is fed to the Y/C Mixer.

239

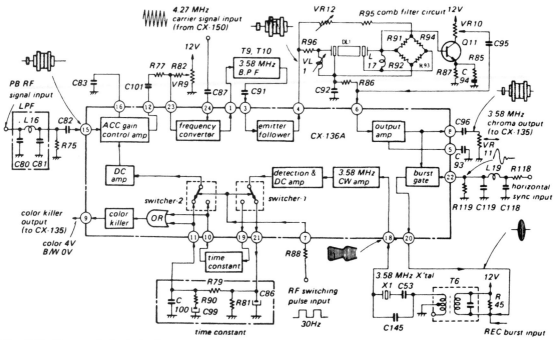

Fig. 7-12. Chroma playback circuit. (Courtesy of Zenith Corp.)

AGC Control Circuit

A burst gate, using a delayed horizontal signal as the gate control, extracts the burst signal. The extracted burst is supplied to the 3.58 MHz crystal filter. This is the same crystal filter circuit used in the record system. As in recording, the burst signal is converted into a 3.58 MHz CW signal of low level. It is amplified to a high enough level to drive the detector circuit.

The 3.58 MHz signal, whose amplitude is proportional to burst, is detected in the detector circuit whose output is fed to a hold circuit via switcher 1. Separate hold circuits are provided for each of the video heads and switching is done by switchers 1 and 2. Using this method the ACC loops become independent for each channel. This results in no signal level difference, in chroma output for the two combined channels, even if there is a big difference in the chroma output levels of the two heads. The RF switching pulse (30Hz) drives switchers 1 and 2. The output of switcher 2 is amplified in a dc amplifier to drive the ACC gain controlled amplifier. The filtered outputs of the time constant are also sent to the color killer circuit via an OR gate. The color killer is set to actuate when the ACC detector output voltage drops below the control range of the loops.

The dc voltages at the output of the time constant circuit decrease so that the ACC amplifier gain increases when the chroma signal is low. When this level decreases below the control range of the ACC, the ACC locks out and the voltages of the time constant circuits drop to about 1 volt. The color killer functions when either of the two channels locks out. The output of the color killer is 4 volts dc in the color mode and zero in the monochrome mode.

CX136A IC Circuit Checks and Tips

The CX136A IC is used for playback chroma processing, encompassing the circuits of playback ACC gain control amplifier, playback color killer circuit and frequency converter to convert 688 kHz playback chroma to the 3.58 MHz chroma output signal. Refer to the block diagram Fig. 7-13 for the correct scope input and output signals along with the proper dc voltages at the pins of the IC.

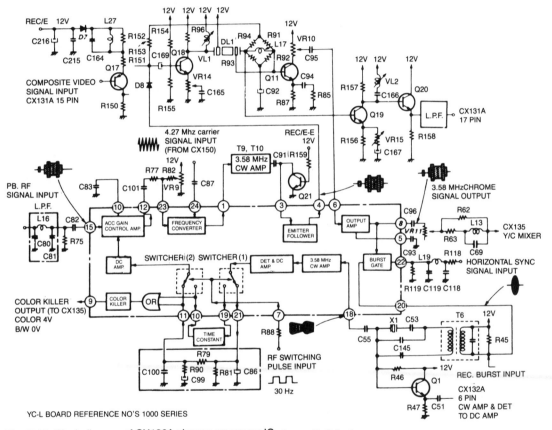

Fig. 7-13. Block diagram of CX136A chroma processor IC. (Courtesy of Zenith Corp.)

Some troubleshooting symptoms that may be caused by this chip circuit are as follows.

- No color
- Color flicker (ACC does not work)
- Carrier leak and/or Cross talk in color signal

For no color check for 3.58 MHz chroma signal (1.9 volts) at pin 8 of the IC. Is color killer output of about 4 volts dc present at pin 9 of IC? Check the 30 Hz switching pulses present at pin 7 of the chip. Check for a chroma signal of 0.3 volt at pins 9 and 10 of the CX135A chip. If all input signals are near normal but output signals are not and the DC voltages seem to be incorrect then the CX136A chip would be a prime suspect.

ZENITH MODEL KR9000 VCR

For normal VCR operation a TV station signal (VHF or UHF) is recorded by the machine and later played back through a television receiver. Thus the input to the KR9000 is at a VHF or UHF frequency, with video chroma and audio modulation.

The Record Mode

Looking at Fig. 7-14, we see the TV input signal appears at the top left of the block diagram. Each input signal is coupled through a splitter, producing two outputs each for VHF and UHF. These tuners operate very similar to tuners in conventional TV receivers.

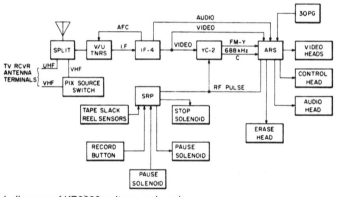

Fig. 7-14. Block diagram of KR9000 units record mode. (Courtesy of Zenith Corp.)

The output of the tuner(s) is at i-f or intermediate frequencies in the 41 to 47 MHz bandwidth. The i-f signal is processed by circuitry on the i-f-4 board, very similar to a TV set operation. The signal is amplified and the three resultant signals developed by detection circuits are the outputs of this i-f functional block. One signal is the recovered video signal, which includes both the luminance and chroma information. Another output is the audio signal portion of the program being processed. A third output is an AFC signal, which is developed by a circuit that senses a change in i-f video carrier frequency caused by tuner oscillator drift. This dc voltage is coupled back to the tuner oscillator circuit, resulting in a continuous correction of oscillator drift, locking the oscillator to the correct frequency for the channel signal received.

As the video head revolves during the *record* process, two magnets on the bottom side pass over fixed coils to create the PG pulses, that are directly related to the speed of the head rotation. These pulses are coupled into the ARS board, where they are compared with a signal developed from the video information, to control and maintain the correct speed of the video head.

The Playback Mode

When the recorded video tape is played back, the heads pick up the information as the tape moves past them. The audio head couples the audio information to the ARS board, for amplification and connection through the CP-3 board to the audio output socket and the RF modulator. Note the block diagram of the playback mode in Fig. 7-15.

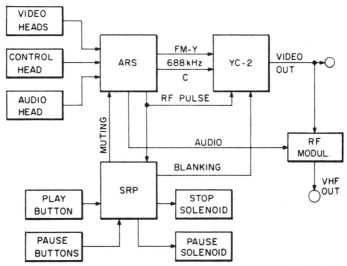

Fig. 7-15. Block diagram of KR9000 units playback mode.

The control head sends the control track signal to the ARS board where it is compared with the 30 PG pulses from the revolving video head for maintaining the correct speed of the video head.

The video heads track the recorded video signal on the tape and the rotating transformer couples this information to the ARS board. The video signal is composed of the FM luminance and 688 kHz chroma signals, which are amplified and separately coupled to the YC-2 board for further processing. This processing recovers the original luminance frequencies as well as the 3.58 MHz Chroma, which are mixed together to form the normal NTSC composite video output that connects to the CP-3 board and the video output socket as well as to the RF modulator.

The audio and video signals are used to modulate either a channel 3 or channel 4 RF carrier, which is coupled to the television receiver for viewing.

Luminance (Video) Playback System

Refer to the block diagram in Fig. 7-16 as we review this video playback system. As the video heads rotate during the *playback* mode they pick up the recorded informa-

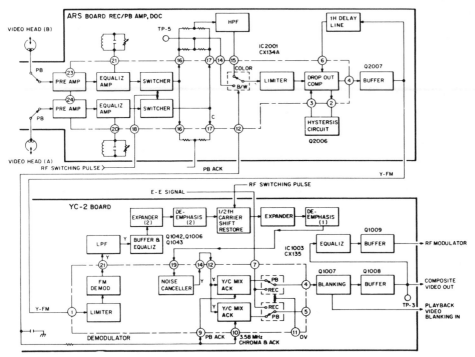

Fig. 7-16. Luminance playback system. <small>(Courtesy of Zenith Corp.)</small>

tion from the video tracks on the video tape. The combined frequency modulated (Y) signal and 688 kHz chroma signals are coupled through the rotating transformer to the ARS board. There is a "chain" of three functional blocks for the output signal of each of the two video heads. The blocks, mostly inside IC2001, are: the preamplifier, the equalizer amplifier, and the switcher circuits. Input to this chain of blocks enter the IC at pins 23 and 24. A 30 Hz signal, the RF switching pulse, at pin 18 switches the processed signals to pins 16 and 17: in time with the active tracking of each head. There are three possible signal outputs from the three center-tapped resistances across pins 16 and 17: monochrome, the luminance portion of the color program, and the chroma (at 688 kHz).

Test Point 5 is the takeoff point for a black and white signal to pin 14 of IC2001 and internally to the B/W position of a switch.

The other switch position is marked "color" and is connected to pin 15. A signal containing both color and B&W can be taken off the top pair of resistors and through the HPF block to pin 15. Only the Y-FM luminance information is connected through the high-pass filter (HPF), because the Y-FM signals are in a high range of frequency by comparison to the chroma frequency. Therefore, no color information goes through to pin 15. Thus the signal that is to be processed further in IC2001 is strictly luminance. The condition of the internal switch connected to pins 14 and 15 is determined by the dc voltage on pin 12. This voltage is the *playback* automatic color killer (PB ACK) signal. If there is no color information in the playback video tape the pin 12 voltage is 0 and the

244

internal switch is in the B/W position. Thus the monochrome signal is coupled from pin 14 through to the Limiter block. If there is color in the video signal the pin 12 voltage is about +4 volts and the internal switch is in the color position and the pin 15 signal is coupled through to the limiter block.

In either case the Luminance signal continues through the *drop out compensator* block and out pin 4 to Q2007, a buffer stage. The buffer output goes in two directions. One is up and to the left through DL2001, a delay line that delays the signal for a period of one horizontal line, about 63 microseconds. The delayed, fed-back signal is coupled to the *drop out compensator* by way of pin 6.

The purpose of the drop out compensator is to insert the 1H (1 horizontal line) of delayed picture information into the main signal in place of dropout, or no information, on a horizontal line during playback.

The Y-FM signal moves out of the ARS board and into the YC-2 board and pin 1 of IC1003. Inside the IC the signal is limited to eliminate amplitude variations before it is coupled into the FM demodulator. The demodulator output, which is the original Y-FM signal as well as the recovered lower frequency luminance signal, appears at pin 21.

A low-pass filter (LPF) selects only the demodulated low frequency luminance and couples it through to a buffer and equalizer block and then upward and on to the right through a chain of functional blocks.

These blocks include expanders and de-emphasis circuits. The emphasis added to the high luminance frequencies for the *record* mode now must be counteracted, otherwise the high frequencies of the playback signal will be distorted, of higher amplitude than the original transmitted TV signal.

The de-emphasized signal returns to IC1003 at pin 19 and inside to the noise canceller block. Out pin 14 and back in pin 12 goes the signal and then into two similar blocks, labelled Y/C mixers. Within these blocks the mixing of the playback luminance and chroma signals takes place and the outputs are connected to the PB (playback) positions of the internal switches coupled to pins 4 and 5. In the playback mode of operation the pin 11 voltage is 0 volts, and the switches are in the PB positions. The playback luminance signal couples out of pin 4, past the blanking block and through the Q1008 buffer, to the *video out* jack and through to the RF modulator.

The blanking transistor, Q1007, is in shunt with the signal path between pin 4 of IC1003 and the buffer stage. During normal operation Q1007 is not conducting and is a high impedance to ground and does not affect the video signal flow. Usually when the VCR is first turned on in the play mode, the tape movement is not up to normal speed. Then the SRP board couples a positive voltage to the base of Q1007, turning it on to saturation. The result is a very low impedance to ground for the output luminance signal, which is thus routed to the ground. Without a video signal into the TV receiver the picture tube is blanked out. When the correct tape speed is reached, the blanking stage is cut off and a normal picture appears on the TV receiver screen.

Clock Timer

A block diagram of the complete clock timer for the Zenith VCR recorder is shown in Fig. 7-17.

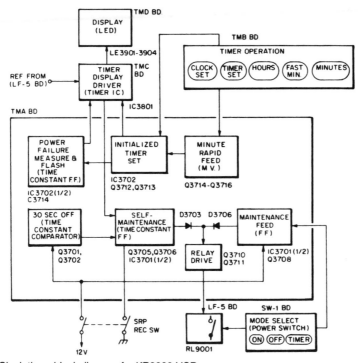

Fig. 7-17. Clock timer block diagram for KR9000 VCR. (Courtesy of Zenith Corp.)

CAPSTAN START CIRCUIT

This circuit is used in the SONY CCD-V110 camcorder shown in Fig. 7-18. The mechanical control CPU controls the operation of the capstan and drum servo. It does this through a series of pulses to ensure rapid starting and stopping of the capstan motor for quick edits and proper handling of the tape. To do this, four outputs are used, (CAP ON) at pin 9, Cap Clockwise/Counterclockwise (CAP CW/CCW) at pin 10, (CAP PULSE) at pin 26, and the Digital Servo Reset (D SERVO RESET) at pin 47. Note the block diagram in Fig. 7-19.

The digital servo reset is a HIGH pulse that is applied to pin 10 of the digital servo, resetting it to provide maximum drive output for quick starting of the servo motors. CAP ON from pin 9 is inverted by Q603, which receives power from the motor power circuit that also provides the power to the loading and mode control motor. If the loading and mode control circuits do not receive power for any reason, the capstan will not be able to operate because it will not be able to get the CAP ON command. The CAP ON command is applied to pin 26 of the capstan driver and enables the driver to drive the motor when pin 26 is high. CAP CW/CCW is inverted by Q513 and applied to pin 27 of

246

Fig. 7-18. Photo of Sony CCD-V110 camcorder. (Courtesy Sony Corp. of America)

the capstan driver. When pin 27 is high, the capstan motor will rotate in a counterclockwise direction and, when low, the capstan motor will rotate in a clockwise direction. Refer to the timing chart shown in Fig. 7-20.

CAP PULSE at pin 26 is inverted by Q517 and applied to pin 23 of the capstan driver. Pin 23 of the capstan driver is the input of the servo drive voltage. During normal play, the output from the servo circuit is applied to pin 23 and drives the motor at the correct speed.

During special effects and braking the power from the CAP PULSE circuit, generated by the mechanical control CPU, takes the place of the capstan servo power in driving the capstan motor. In going from the STOP mode to the PLAYBACK mode, it is desirable to have the capstan servo quickly lock up. This is achieved by the mechanical control CPU, which causes the CAP ON command to go low at the positive transition of the RF switching pulse that occurs during the F1 field. The digital servo reset pulse goes high 3 ms later for 180 μs. This pulse resets the digital servo to provide a maxi-

247

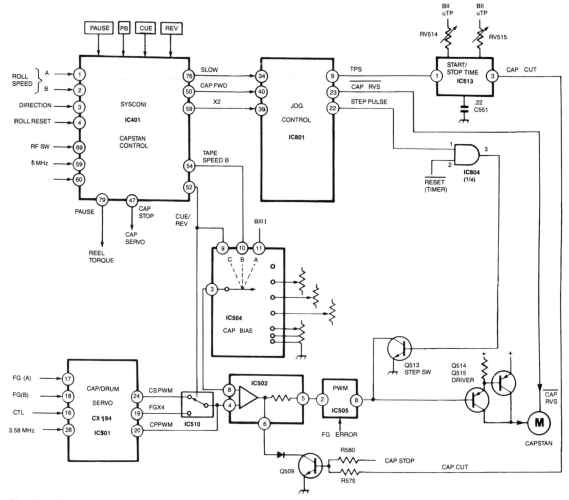

Fig. 7-19. Block diagram of capstan control—fast (Courtesy of Sony Corp. of America)

mum output to quickly bring the capstan motor to speed. During this time, the CAP CW/CCW output remained low to move the tape in the *forward* direction, which corresponds to a CCW rotation of the capstan motor.

In going from the STOP to the FAST FORWARD direction, care must be taken to avoid quickly starting the motor and thereby possibly damaging the tape. To prevent tape damage, the CAP ON pulse does not go low immediately. When the FAST FORWARD mode is first entered, it goes low for 10 ms and high for 37.5 ms before going low again for another 10 ms and high for 37 ms. The width of the positive duration of the CAP ON pulse gradually decreases until, after approximately 2 seconds, the CAP ON pulse will remain low, applying full power to the capstan motor and quickly moving the tape.

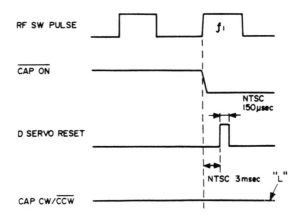

STOP → PB Capstan Start Timing Chart

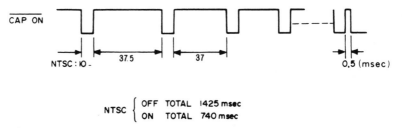

$$NTSC \begin{cases} OFF \ TOTAL \ 1425 \ msec \\ ON \ TOTAL \ 740 \ msec \end{cases}$$

Fig. 7-20. Stop—Fast Forward timing chart. (Courtesy of Sony Corp. of America)

In going from the FORWARD to the STOP mode, it is desirable to bring the tape quickly to a stop. Note the waveforms in Fig. 7-21. To do this, the mechanical control CPU generates a braking action at the positive transition of the RF switching pulse in the F1 field. This braking action is produced by holding the CAP ON signal low so that the capstan motor can continue to receive power. Power is applied by a CAP PULSE output from the mechanical control CPU that is inverted and goes High for 4 ms to provide power to the capstan servo. This power drives the motor in the reverse direction, as CAP CW/CCW is high during this period, causing the motor to turn in the reverse direction. This quickly brings the motor to a stop. After 1,000 ms, when the motor has fully stopped and the CAP ON signal has gone high, disabling the capstan motor, the main brake is energized.

In going from REWIND to STOP, the action is similar, as it is desirable to bring the tape to a quick stop. Note the capstan timing in Fig. 7-22. During this time, while CAP ON is still low, CAP CW/CCW goes low for 50 ms so that the power driving the motor now drives the motor in the reverse direction, quickly stopping it. During this time, the main brake has also gone high, energizing the main brake, and the CAP PULSE output remains low as this drive is not needed.

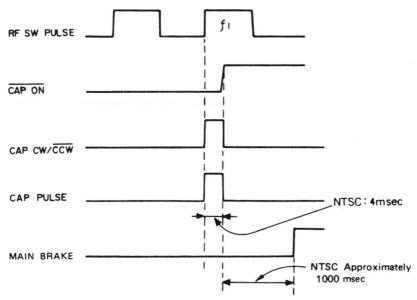

Fig. 7-21. Forward—Stop capstan timing. (Courtesy of Sony Corp. of America)

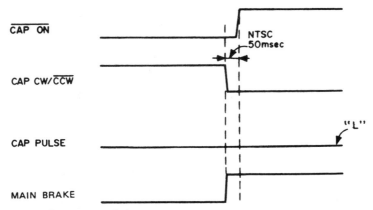

Fig. 7-22. Rewind—Stop capstan timing. (Courtesy of Sony Corp. of America)

Capstan Motor Drive

The CAP ERROR and CAP BIAS voltages are applied to a comparator within the capstan motor driver to produce the driving voltage for the capstan motor. This voltage is amplified within IC502 and applied to a PWM that uses a sawtooth from TF4 at IC502/pin 17 to produce a PWM output at IC502/pin 16. Note the circuit in Fig. 7-23.

The PWM output is amplified by Q503 and applied to the current amplifier Q502. The current pulses are filtered and the resultant dc voltage is applied to IC502/pin 15. Here it is applied to a current limiter, which removes bias from the capstan servo voltage amplifier if the voltage at pin 15 exceeds CAP Vcc. This voltage is the drive current for the capstan motor. It is the power source for the current sources at pins 10,

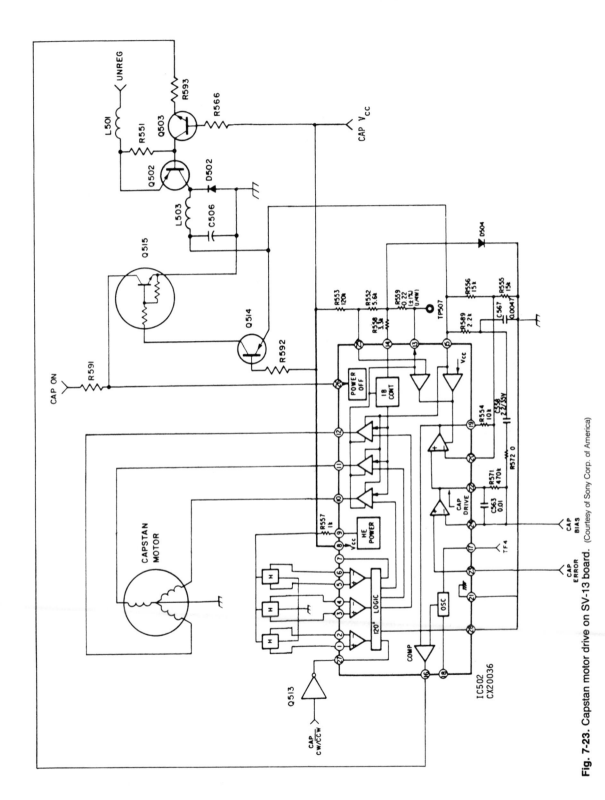

Fig. 7-23. Capstan motor drive on SV-13 board. (Courtesy of Sony Corp. of America)

251

11, and 12, which provide the current for the capstan motor. The phase of the current through the capstan motor windings is controlled by the output of three Hall devices mounted on the capstan motor. These outputs are amplified within IC502 and applied to a 120 degree logic block that determines which of the three current amplifiers is *on* at any particular time.

Since the capstan motor can turn in both the clockwise and counterclockwise directions, a command from the mechanism CPU (CAP CW/CCW) is applied to IC502/pin 27, after going through an inverter, Q513. If pin 27, is high, which in this case is above 0.7 V, the capstan motor will turn in the counterclockwise direction. This is the normal mode of operation for the capstan motor. During REVERSE or REVIEW modes, pin 27 will be low and the capstan motor will turn in the clockwise direction, moving tape in the opposite direction. The direction of the capstan motor is changed by controlling the phase relationship of the Hall device output, which controls the current drivers.

The capstan driver and capstan motors are protected in many ways from excessive drive current. To turn the unit on, CAP ON is produced by the mechanism control that makes pin 26 high, enabling the capstan motor driver. If the drive voltage at IC502/pin 15 becomes excessive and exceeds CAP Vcc, Q514 will turn on and apply a high voltage to the base of Q515, turning it on and pulling the voltage at IC502/pin 26 low. This turns off the capstan motor driver.

MONOSTABLE MULTIVIBRATOR CIRCUIT

For the best picture during the PAUSE mode, it is desirable to stop the tape at the beginning of the F1 field. To do this, the mechanical control CPU uses a monostable multivibrator to time the position of the tape. The monostable multivibrator produces a 5 ms negative pulse at the negative transition of the RF switching pulses. Note the timing waveforms shown in Fig. 7-24. These negative pulses indicate the position of the F2 and F4 fields of the information. The output of the monostable multivibrator is applied to the mechanical control CPU, which will enter the STOP mode by starting the BRAKE command rising 8 ms after the pulse that occurred during the F4 field. In this way, the brake pulse reaches its maximum, entering the STOP mode at the beginning of the F1 field. A control is provided to adjust the width of the monostable multivibrator pulse to precisely control the position that the tape stops.

The monostable multivibrator circuit shown in Fig. 7-23, is a part of the matrix for the mechanical control CPU. This matrix is also used to read the Tape Top, Tape End, AFT Lock and Mode Switch position inputs. The RF switching pulses are differentiated by C608 and R642, which produces short pulses at the cathode of D603. The negative pulses are coupled through C611 and trigger the monostable multivibrator, producing a high output at Q626s collector. The duration of this high is determined by the time constants of R643, C611, R640, and VR501. VR501 is the *still* adjustment. This is not a customer adjustment. It is an internal adjustment, and it will precisely position where on the tape the unit will stop. The high pulse produced by the monostable multivibrator is

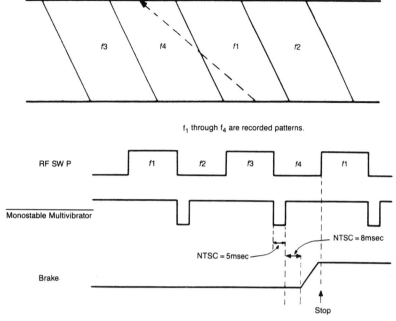

f_1 through f_4 are recorded patterns.

Cf: Stopping at trace of f_1 and f_4 is made in accordance with the CX20035 JOG VD generation method.

Fig. 7-24. Timing waveforms. (Courtesy of Sony Corp. of America)

inverted by Q627 and applied through Q602, when it is turned on by the MATRIX OUT 1 command, to MATRIX IN 4 where it is sensed by the mechanical control CPU. A photo of the SONY model CCD-V4 Camcorder is shown in Fig. 7-25.

Luminance Encoding

The luminance signal from the aperture correction circuit is fully processed to contain the correct luminance information for color balance and detail. In addition, the detail has been enhanced by the aperture correction circuit. It now must be brought up to the NTSC standards by adjusting the pedestal, limiting peak level, and adding sync. This is done in the circuits contained in IC703. Refer to the circuit shown in Fig. 7-26.

"Y" is first applied to a gain control amp whose gain is fixed by the dc voltage at IC703/pin 29. Upon leaving the gain control amp, "Y" is blanked by BLK entering IC703/pin 5; and the pedestal level is set in the pedestal circuit. The pedestal level is controlled by RV716 at pin 30. After the pedestal is properly set, the white clip circuit, controlled by RV717 at pin 31, ensures that the luminance signal does not exceed the maximum value. "Y" is then coupled from IC703/pin 36 to IC703/pin 37 through C744. On reentering IC703, "Y" takes two paths. The first path is to a mixer where sync is added. Composite sync from the sync generator enters IC703/pin 3. It is amplified by a gain control amplifier, whose gain is determined by the fixed voltage at pin 32. Once the sync is added, "Y" is then applied via C743, the muting switch (always closed), and exits pin 47 to ultimately appear in the camcorder viewfinder.

Fig. 7-25. Photo of SONY CCD-V4 camcorder. (Courtesy of Sony Corp. of America)

The second path is to a fader amplifier. The gain of the fader amplifier is controlled by the voltage at pin 33. This same voltage will be used to control the level of the chroma signal in another circuit. The "Y" output of the fader circuit is mixed with the sync, chroma, and burst before exiting IC703/pin 44. C742 then couples the "Y" signal to the video processor and camera output. Sync and burst are added to "Y" after the fader to prevent loss of servo control in the VCR. If this is not done sync and burst will not be at the correct level during fader, and the VCR servos will not be locked to them.

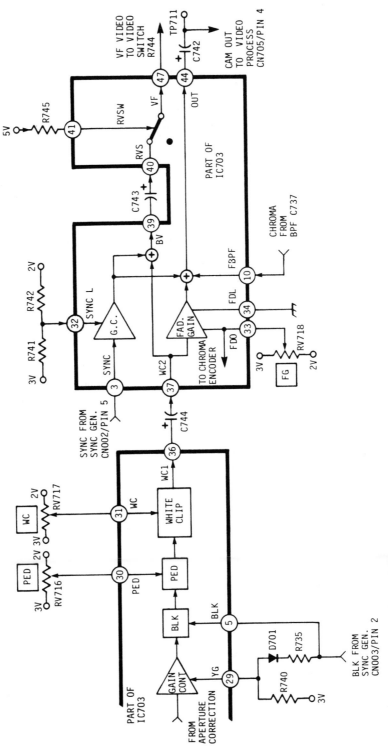

Fig. 7-26. Luminance encoding on ME-1 board. (Courtesy Sony Corp. of America)

Chroma Encoding

The color signals from the matrix circuit do not need further processing other than low-pass filtering. The color signals R-Y and B-Y enter IC703/pin 13 and 14, respectively. They are clamped by the CLP3 and applied to two amplifiers with the dummy dc voltage as the reference. The output of these amps is connected to the chroma modulators where the 3.58 MHz chroma carriers (which differ by 90 degrees) are modulated. The chroma encoding circuit is shown in Fig. 7-27.

The color subcarrier is derived from the 4FSC signal that enters pin 8. This signal is divided by 4 to produce the 3.58 MHz chroma subcarrier and phase shifted in a phase shift block to produce the two carriers that differ in phase by 90 degrees. This phase shift is fixed, although it can be made to vary by changing the voltage at pin 1 and pin 24 in non-NTSC formats. The color subcarriers are applied to the appropriate chroma balanced modulators where the R-Y and B-Y signals control the output of the 3.58 MHz signal. Voltages at pins 25 and 26 control the modulator balance so that when no color signals are present (during a black scene), the outputs of the modulators are mixed and applied to a fader circuit controlled by the same voltage that controls the "Y" fader gain. After the fader, burst is added to the chroma signal.

Burst is produced by taking two different phase outputs of the 3.58 MHz subcarrier from the phase shift circuit and applying them to balanced modulators. The gain of one of the modulators is controlled by the voltage at pin 28. This is the hue adjustment of the burst. It varies the phase of the burst signal and, therefore, the hue of the color signal. This is done by varying the gain of one of the two different phase signals so that the resultant phase can very slightly. The burst goes through a gain control amp that produces a constant amplitude burst signal (determined by the voltage at pin 27). It is switched by the burst flag at pin 2 to produce the burst signal. Burst is added to the chroma signal that exits pin 11, goes through a bandpass filter, and reenters IC703/pin 10 where it is combined with the luminance signal.

SONY SUPER BETA VCR

The following circuit information and troubleshooting hints are for the Sony Super Beta VCR machines.

Head Switching

With the use of 4 heads, a single clogged or defective head is easy to identify, as well as which head is used or not used in which mode. However, if noise in the picture occurs in one of two modes that uses the same heads, the problem may be in the head switching circuitry that selects the heads.

The head switching circuitry chooses a pair of heads with the information from system control. In special effects modes, information also comes from the drop out compensator to identify a loss of RF from that active head, and uses it to switch to another head that is beginning to pick up RF. Refer to Fig. 7-28 for the switching circuit.

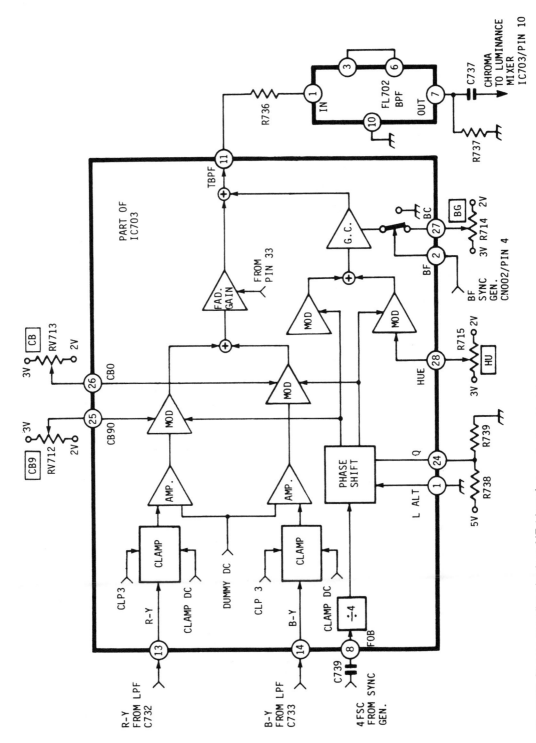

Fig. 7-27. Chroma encoding circuit on ME-1 board. (Courtesy Sony Corp. of America)

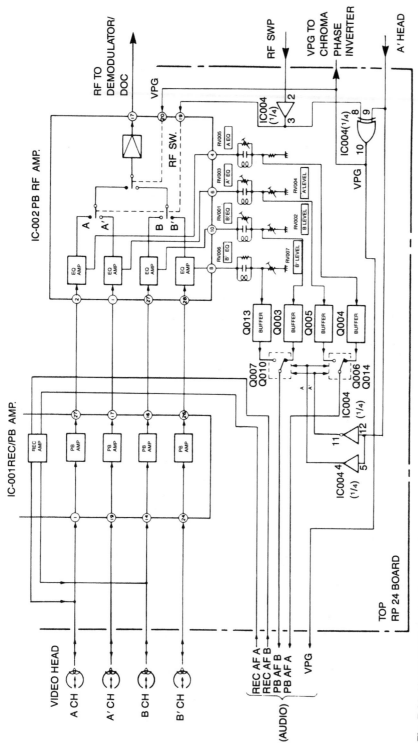

Fig. 7-28. Head switching circuits. (Courtesy Sony Corp. of America)

During playback two heads are selected at a time. They are chosen by switches in IC002/pin 20. The 30 Hz RF switching pulse comes from the head drum assembly and is used to select the heads from either side of a video head disc that are in contact with the tape. For example, when the RF switching pulse is high either B or A′ heads at one side of the video head disc are available. When the RF switching pulse is Low, A or B′ heads on the other side are available. The *video pulse generating* (VPG) signal is the other signal needed to select a pair of heads. This pulse comes from an EXCLUSIVE OR gate, and along with the RF switching pulse at IC002/pin 19 is used to select between the two adjacent heads on the same side of the 4 head disc.

The EXCLUSIVE OR gate makes the VPG head selection signal from an RF switching pulse and an A′ HEAD signal. The A′ HEAD signal in turn comes from system control and the DOC circuit to select the correct heads for lowest noise signal in special effects.

The head switching signals chart (Fig. 7-29), may be used with the head switching schematic to determine which heads are used in which special effects modes of operation. In normal playback the A and B heads are used. They are chosen by IC002 with an

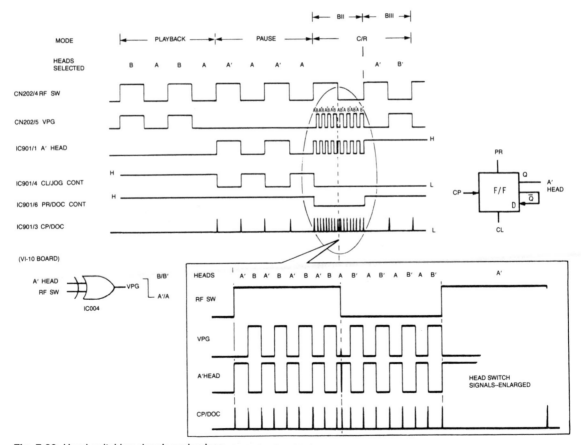

Fig. 7-29. Head switching signals and pulses. (Courtesy Sony Corp. of America)

259

RF switching signal at pin 20 and a duplicate RF switching pulse at IC002/pin 19, the VPG signal input. The VPG, which is only an RF switching pulse during normal PB, comes from IC004/pin 10. When the A' head signal at one input of IC004/pin 9 is low, the other input, an RF switching pulse at IC004/pin 8 is allowed to output at IC004/pin 10 to become the VPG signal. Therefore, in normal PB, the A' head signal is low to produce a duplicate RF switching pulse signal at VPG to select the A and B heads.

When pause is initiated, a 30 Hz square wave in phase with the RFSW pulse appears as the A' head signal at IC004/pin 9. The A' head signal is joined with an RF SW pulse to the inputs of EXCLUSIVE OR gate IC004. The in-phase square waves produce a constant low at the VPG output during pause. The VPG signal is applied to IC002/pin 20. This is one of two signals used to select a pair of video heads. The other input IC002/pin 19 is the RF SW pulse. With a low at IC002/pin 20 and a RF switching pulse at IC002/pin 19, heads A and A' are selected in pause for a noiseless still picture.

Channel Change

Channel changes are made with the front panel jog dial through system control 1, IC401. The routine of communications is the same as with POWER ON. Note the circuit in Fig. 7-30. When the jog dial is turned, ROL pulses through IC401 and produces request pulses to the timer control IC that returns with serial clock pulses to permit serial data to enter the timer control chip. This updates the clock or channel display. The central timer control chip in turn sends out channel serial data and clock pulses to the tuner control IC. When the jog dial has stopped turning the last channel number is held by the tuner control IC and station memory information for that channel is accessed. Before the station is tuned in, the tuner control IC sends another request pulse to the timer control along with channel number data for comparison to the display number. If they agree, verification data is sent to the tuner IC. Then the tuner control IC sends data using the clock and latch line up into the VHF tuner to make a VC voltage that is used to select the station. Fine tuning for that station is the next step only if proof of station existence is presented to the tuner control IC in the form of horizontal sync pulses at IC101/pin 12. Fine frequency tuning is achieved by monitoring differential AFT signals arriving at tuner control IC101/pins 40 and 41. When the voltages at these pins are equal, the station is fine tuned in and no more correction data is delivered to the VHF tuner.

Time Change

Time change, like channel change is also initiated with the Jog Dial. It passes through system control 1, IC401 in the same manner as the channel commands. What determines whether the Jog Dial is used for channel change or time change is the time-set push button in the keyboard matrix, tied to the timer control IC101. If the time set or CLOCK buttons are not pressed, the Jog Dial is used to change stations.

The clock or time changes are made from the front panel keyboard directly to timer control IC101 and it's display update to reflect the changes stored internally. Also

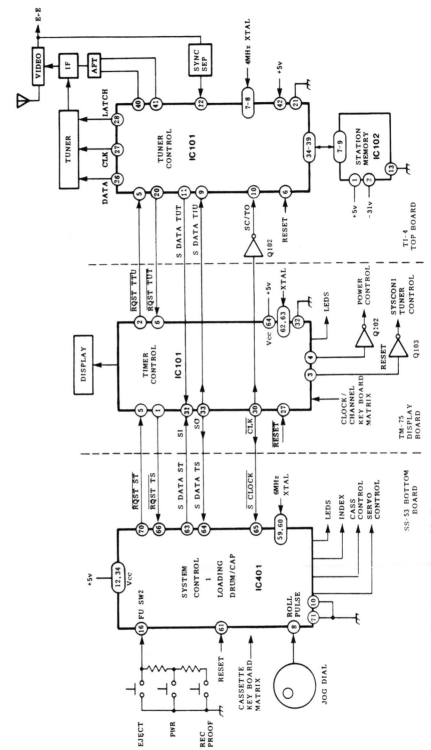

Fig. 7-30. System control — timer and tuner block. (Courtesy Sony Corp. of America)

261

stored within the timer control IC101 are the times and the days of the week for the timer activated recording feature, for up to eight programs within a three week period. The channels for the eight programs, however, are stored in the station memory, IC102 accessed through the tuner control IC101.

Loading/Threading—Sony Super Beta

The loading motor is used to draw the Betamax cassette in, lower, and thread the tape around the drum. This operation is centered around system control IC401. Refer to Fig. 7-31. After power is turned ON, inserting the cassette part-way closes a cassette switch in the right hand side of the mechanism compartment. This is electrically shown as the CASSETTE ON, or CASSETTE IN switch, tied into the resistance network entering IC401/pin 13. When this closure is detected, first the pinch roller solenoid is engaged by using IC401/pins 25 and 26. Both pins go high initially to engage the pinch roller solenoid plunger, then pin 25 returns low. Pin 25 is only used to supply the very high amount of current to the pinch solenoid to draw the plunger in initially. Thereafter, the high at pin 26 continues to hold the plunger in while the loading motor is activated, drawing the cassette in, and lowering it into position on the reel tables. When the cassette has fully lowered onto the cassette tables, a cassette down switch closes in the cassette compartment. This is electrically shown as the middle switch at IC401/pin 13. When the switch is closed the cassette has lowered into position and the pinch roller solenoid disengages. This changes the loading motor gearing assembly, permitting the loading motor to now thread the tape around the drum. When the threading ring reaches the end of the stop position, the thread end microswitch closes, grounding IC401/pin 13, completing the entire sequence and stopping the loading motor.

The EJECT button reverses this entire procedure. When IC401/pin 16 is momentarily grounded with the EJECT button, the voltages on pin 57 and 58 reverse, permitting unthreading. At the very end of unthreading, when the threading ring has reached the stop, the top or right hand side tape end sensor coil is moved mechanically by the threading ring toward a piece of metal. The metal lowers the Q of the resonate circuit producing a low at IC401/pin 24 marking the conclusion of threading.

The pinch roller solenoid plunger now engages and unloading begins. In concludes when the CASSETTE IN switch at IC401/pin 13 opens.

Playback and Record

Main system control IC401 interprets the front panel cassette command keys and initiates that function by sequentially controlling the video processing, capstan servo, and drum servos. During record and playback the time counter display is periodically updated with elapsed time information via timer IC101. Note the circuit in Fig. 7-32. The Record/PB control operation will now be explained. The front panel cassette control push buttons are arranged in a key matrix network as one input to system control 1, IC-401. Usually, key matrix networks have a pulse periodically scanning the keys and when the key is pressed or closed, the pulse is returned to one of the inputs. Although

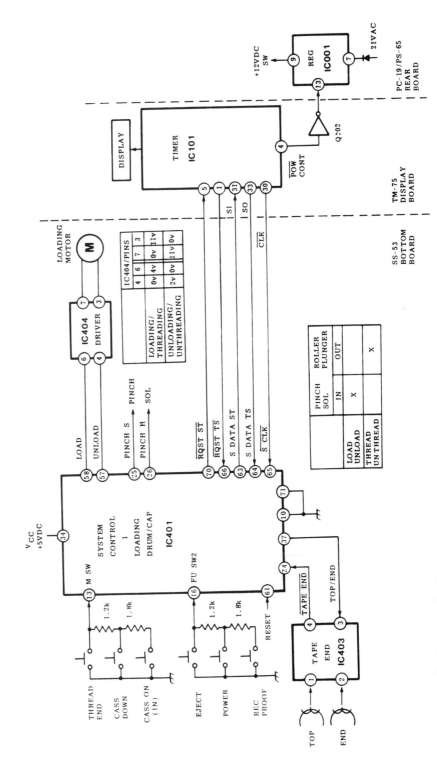

Fig. 7-31. Loading/threading circuitry. (Courtesy Sony Corp. of America)

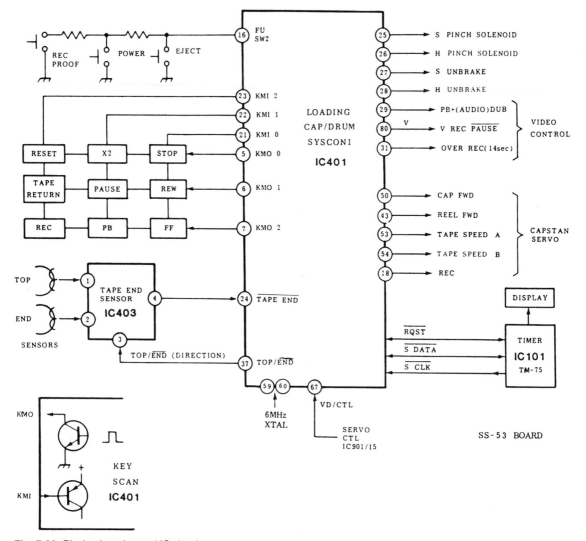

Fig. 7-32. Playback and record IC circuit. (Courtesy Sony Corp. of America)

this is how it works, there are no pulses that leave the output ports. The reason can be seen by using the simplified diagram below the key matrix in Fig. 7-32. Examining the top transistor, you can see a pulse coming into the base. However, the collector is open circuited. This is the KMO output. When a push button is pressed the KMO output is coupled to the KMI input. This completes a B+ path for the open collector transistor, permitting the pulses to pass through. Therefore, at the output port you would see 0 Vdc and no pulse. At the KMI input port there would be a positive voltage and no signal. Key scan pulses are derived from the 6 MHz clock at IC401/pin 59 and 60.

Whether playback or record is selected using the cassette push buttons, IC401 checks for completion of threading by looking for a ground at pin 13. If RECORD was selected, the record proof switch at IC401/pin 16 is checked as well. The tape will be ejected now if the tape's record tab was removed. When these conditions at IC401/pin 13 and 16 are satisfied the pinch roller and unbraking solenoid are engaged, followed by control to start the capstan servo for tape handling and transport. Video Playback is enabled with a high at IC401/pin 29, while video record is a high at IC401/pin 80. During the first 14 seconds of a video edit recording, pin 31 goes high to increase the record current (luminance signal) to record over whatever is on the tape initially.

While the tape is moving, VD pulses in record or CTL pulses in playback enter pin 67 of the IC. They are interpreted and they appear as request and serial data signals at timer IC101 once every second. This is used to update the real (elapse) time counter display. At the end of the tape, metal foil passes across the tape end coil end sensor. Since the coil sensor is the inductor in the L-C Resonate circuit, the Q drops, and therefore its output level. This change is detected by the tape end sensor IC403, producing a sharp low into IC401 pin 24 terminating record or playback. After tape end, IC401 rewinds the tape to the beginning, identified by another metal foil there. This tape end sensor coil on the other side of the cassette compartment, switches ON during rewind, to stop the tape.

SONY 8 MM PCM EV-5700U RECORDER

The Sony 8 mm video format uses the helical scan system for video recording that is used in virtually all video recorders. This system produces the high video head speed with a relatively low tape speed, that is essential for recording the high frequency video information. But the 8 mm video format has been designed with many additional features, and for this reason the tape is divided into four areas for recording information. Two of these areas are scanned by the rotating video heads producing the slanted tracks characteristic of all video recorders. The other two areas are conventional longitudinal tracks. Note the 8 mm format shown in Fig. 7-33.

The first and largest area contains the video, both luminance and chroma, the FM mono audio signal, and the tracking signal, which is used in this format to maintain tape tracking. In the 8 mm format only this portion of the tape must be recorded, the remaining three areas are optional and used for special features. All the information needed is found in this portion of the tape, recorded by the rotating video heads; the video, the audio, and the tracking signal, which replaced the CLT signal.

As in a Betamax two video heads are found, designated head one and head two, and they are distinguished from each other by a ± 10 degree azimuth difference to reduce cross talk between the adjacent channels. The width of the tape track is 20.5 microns in the NTSC format.

The second area of information extends for 1.25 mm below the video track. This area is for the PCM audio and tracking signal, and is reserved for the recording of stereo PCM encoded audio. It is not a requirement of the 8 mm format that all units play this PCM audio, but this portion of the tape must be available on all 8 mm recordings. A tape, which has been recorded with the PCM audio information will also have the FM

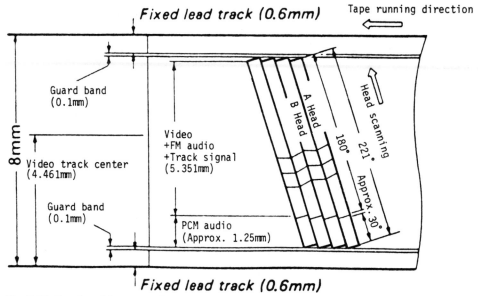

Fig. 7-33. Drawing of 8 mm tape format. (Courtesy Sony Corp. of America)

audio mixed with the video signal. Thus, this tape can be played on a unit without the PCM capability which will produce sound from the FM audio. It is this compatibility between all machines, both present and future, that makes the 8 mm format so attractive. In addition to the PCM audio signal, the tracking signal is also recorded in this area of the tape.

An optional cue track, 0.6 mm wide is located at the top of the tape. This cue track is separated from the video tracks by a 0.1 mm guard band. This cue track is optional, like the PCM audio area. It may be used on future machines for the recording of editing information. Like the PCM signal, this cannot be recorded without also recording the full video, FM audio and tracking signal in the scanning portion of the tape so that any tape made with, or without, a cue signal can be played on any 8 mm machine.

There is also a longitudinal audio track at the bottom of the tape. This track is 0.6 mm wide and, like the cue track, is separated by a 0.1 mm guard band. This longitudinal audio track is optional. A recorder that uses this track must still record using the FM audio mixed in with the video and tracking signal on the tape, and no unit can be made using only this longitudinal audio track, as it will not be able to play 8 mm tapes made on 8 mm units without it.

PCM Audio Signal Record Mode

The 10 bit serial data from the A/D converter enters IC101/pin 16. It is synchronized with the bit clock and word clock signals at IC101/pins 17 and 18 respectively. From pin 16, it enters the block labeled 8 to 10 Bit Conversion. (In the record mode, it is actually 10 to 8 bit conversion). It then exits IV101/pin 9 and enters IC102/pin 46. The block diagram is shown in Fig. 7-34.

266

The timing for this is the transfer clock output from IC101/pin 4. Inside IC102, the serial data enters the data interleave block where it is converted from serial to parallel data. In then enters the RAM port controller block.

The RAM port controller controls movement of the data in and out of the two 16K RAMS. The RAMs are used to perform cross interleaving and the cyclic redundancy check code. At the same time, ID data from IC154 is added. The RAMs also hold the data, or compress it, until the head that corresponds with the data comes into contact with the tape again.

From the CRCC block, the data enters the FM modulator where the high's and low's are converted to FM RF. It then exits IC102/pin 8, enters IC103/pin 10, is buffered, and exits IC103/pin 12 as REC PCM RF.

PCM Audio Signal Playback Mode

In the playback mode, PB PCM RF enters IC103/pin 20 where it goes through a low-pass filter and a high-pass filter. This is the equivalent of a bandpass filter. It then takes two paths. One to the D input of a flip-flop and one into a PLL (phase-lock loop) block. Again, refer to Fig. 7-34. This results in two signals being generated. The signal exiting IC103/pin 10 contains the digital PCM playback data, but is still bi-phase modulated. At the same time, the signal exiting IC103/pin 8 is a transfer clock that is used to clock the PCM playback data into IC102.

Inside IC102, the data is demodulated and CRCC detection/correction is performed. From the CRCC block, the data enters the RAM port controller. The RAM port controller takes the serial data, converts it to parallel data, and alternately stores it in the two 16K RAM's. Now the data can be de-interleaved. At the same time the ID data is taken off and sent to the ID data processor IC154. From the RAM port controller, the data is sent to the data interleave block, where it is converted from serial to parallel. If an error that cannot be corrected is present, the error flag output goes high at pin 2. The serial data then exits IC102/pin 46 in synchronization with the transfer clock input at IC102/pin 3.

At IC101, the serial data enters the block where 8 to 10 bit conversion takes place. From there it goes into the block marked interpolation. At this point, if the error flag at IC101/pin 5 is high, an uncorrectable error has occurred. In this case, the interpolation block tries to reconstruct the data by interpolating or by using the data from the previous field. It then exists the IC101/pin 16 in synchronization with the bit clock and word clock at IC101/pins 17 and 18.

The master clock generator for the PCM process is IC104. The output at IC104/pin 8 is an 11.6 MHz master clock signal.

IC001 is called the feature CPU. This is another name for system control. Its function is to interface the PCM block with the other blocks in the EV-S700U.

IC151, IC152, and IC153 are used only in the multi-PCM mode. Their purpose is to shift the timing of the RF SW PULSE, SREF, and Flying Erase head so that they occur in accordance with the multi-PCM track being used.

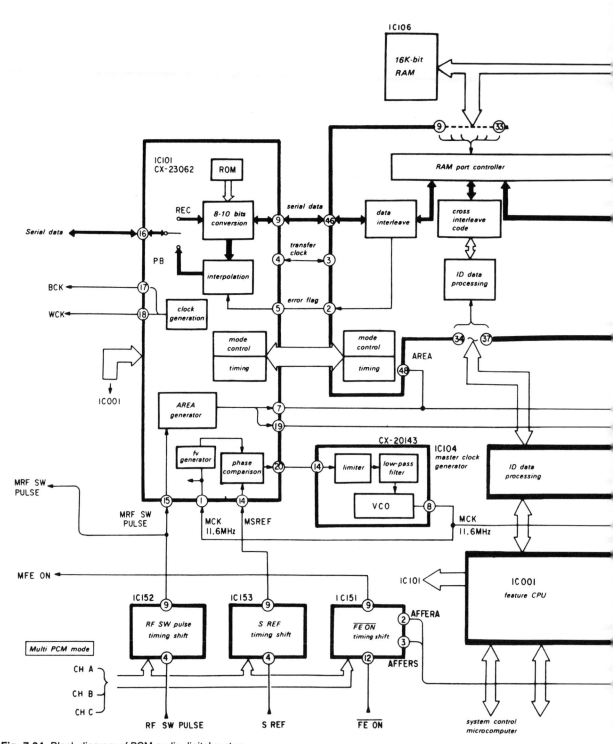

Fig. 7-34. Block diagram of PCM audio digital system. (Courtesy Sony Corp. of America)

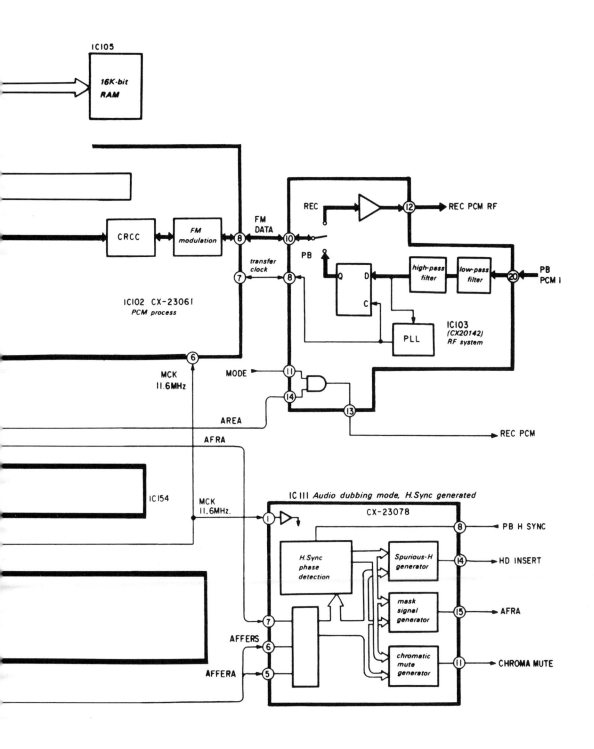

IC111 is used only in the Audio Dub mode. Its purpose is to blank certain parts of the playback picture that would normally become disturbed during the audio dubbing process.

A/D-D/A Converter Record Mode

In the EV-S700U the A/D and D/A process are both done with IC604, located on the PC-14A board. The circuit diagram is shown in Fig. 7-35.

Record Mode

In the Record mode, IC604 is responsible for performing the sample and hold function and also converting the PAM signals into two's complement form.

The signal flow will be explained using the left channel only. The left channel audio signal enters IC604/pin 10, where it is amplified by a gain of 4 by the L CH Aperture amplifier. From pin 12 the signal proceeds through C322, R322, RV304, and C323 into pin 13. The amplitude has now been reduced by approximately ½. From pin 13 the signal is then sent to the L CH Sample/Hold amplifier. C613 and the op amp between pins 4 and 6 form an integrator. This takes the output from the Sample/Hold amplifier and forms a PAM signal at pin 6. Note that the PAM signal at pin 6 contains both left and right channel information. This is because the switches inside of IC604 have alternately selected, or multiplexed, the left and right channels going into the integrator.

The PAM signal at pin 6 is sent to two voltage comparators each having a different reference voltage. The outputs from the two comparators gate a counter. IC604 then converts the output of the counter into two's complement. The signal is then output at pin 28. The timing of the output data is controlled by the BCK, bit clock, and WCK, word clock, signals at pins 26 and 27 respectively.

A/D-D/A Converter Playback Mode

In the playback mode, data in two's complement enters IC604/pin 28. The data transfer to IC604 from IC101 is controlled by WCK, word clock, at pin 27 and BCK, bit clock, at pin 26. Inside IC604 the data is separated into the left and right channel information and converted into analog form. The left channel is output at pin 12 and the right channel is output at pin 17.

Segment Ext. Block Diagram

In order to identify the multi-PCM mode, a 230 kHz carrier is mixed with the FM PCM data in the Record mode. In the playback mode, when this 230 kHz carrier is detected, the EV-S700U knows that a multi-PCM recording has been made. It can also compare the location of the 230 kHz carrier to the RF switching pulse and indicate which track has PCM information on it. Refer to the diagram in Fig. 7-36.

The segment exist circuit consists of IC801 and IC802, which are tone decoders. A tone decoder is an IC that is able to detect a particular frequency. The components

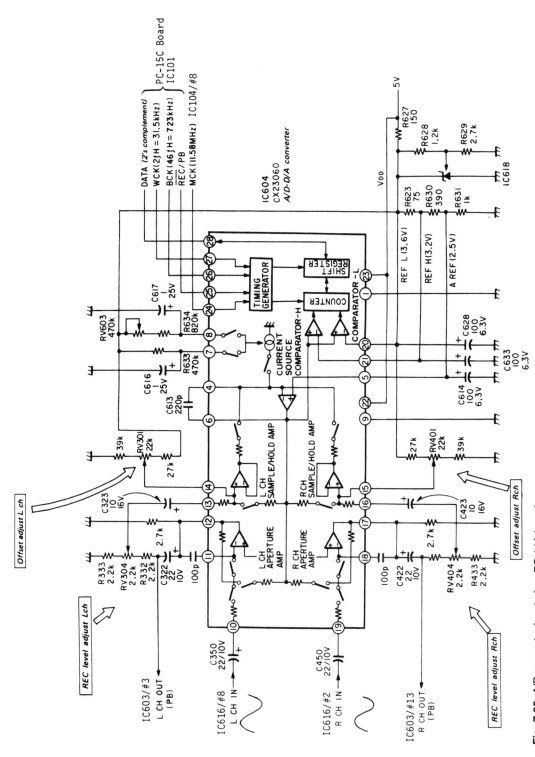

Fig. 7-35. A/D converter located on PC-14A board. (Courtesy Sony Corp. of America)

271

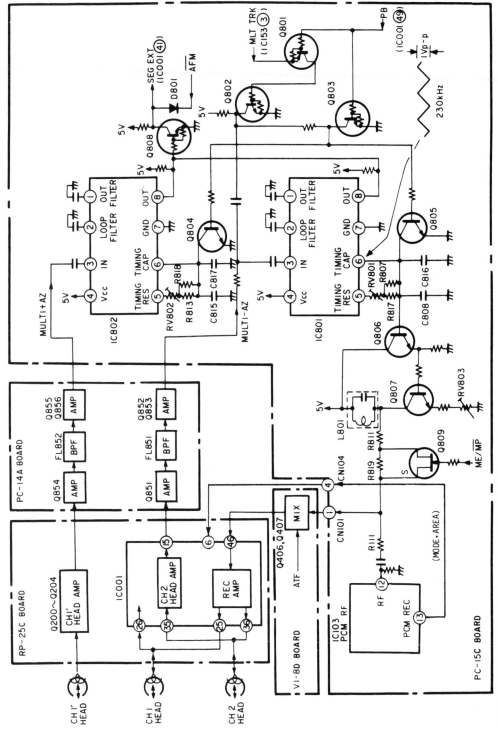

Fig. 7-36. SEG EXT block diagram located on PC-15C board. (Courtesy Sony Corp. of America)

associated with pin 5 and pin 6 determine an internal oscillator frequency. When the input at pin 3 equals the oscillator frequency, the output at pin 8, which is normally high, goes low.

In the normal Record mode, the PB signal and the MLT TRK signal are both low. This turns Q801, Q802, and Q803 all OFF. The 5 volts from Q802/C can now turn ON Q804 and Q805. This prevents the oscillator in either IC801 and IC802 from turning on. Therefore, the 230 kHz carrier is not recorded in the normal Record mode.

In the multi-PCM Record mode, the PB signal is low and the MLT TRK is high. This turns Q803 OFF and Q801 and Q802 ON. Now Q805 is OFF and oscillations are present at IC801/pins 5 and 6. The output is a 1 Vp-p 230 kHz triangular wave. This triangular wave is buffered by Q806 and amplified by Q807. The filtering network on Q807/C smoothes the triangular wave into a sine wave.

The sine wave passes through R811 and R819 where it is mixed with the REC PCM RF signal and proceeds to the record amplifiers.

Q809 is an FET that alters the amplitude of the segment exist signal depending on whether metal oxide or metal evaporated tape is used. Note that even though Q804 is turned OFF, the oscillator on IC804/pins 5 and 6 is not used for anything in the Record mode.

In the Playback mode, the PB signal goes high. This turns Q801 and Q802 OFF and Q803 is ON. It makes no difference what state the MLT TRK signal is on Q801/E. With these conditions, Q804 and Q805 are both turned OFF allowing the oscillators to run on pins 5 and 6 of IC801 and IC802.

During Playback, the RF signal from the channel 2 head enters IC001/pin 35 on the RP-25C board and is amplified by the head amp inside the IC. It then exits at pin 15 where it enters the PC-14A board. Here it is amplified by Q851. It then goes through a bandpass filter, FL851. This bandpass filter allows only the 230 kHz signal to pass. It is then amplified by Q852 and Q853. From here it enters the PC-15C board and IC801/pin 3. If the 230 kHz is detected, the output at IC801/pin 8 goes low. This turns Q808 OFF and puts a high on the collector indicating that multi-PCM exists on the tape. IC001, system control, can then compare the timing of these high pulse(s) to RF SW P and correctly indicate which multi-PCM track has PCM recorded on it.

Notice also that the output from the channel 1' head amp is fed to IC802/pin 3. The channel 1' head is used so that the multi-PCM signal can be detected in the cue and review modes and also in pause.

8

Introduction
To Microcomputer Systems
and Microprocessors

The microprocessor is a large, complex integrated circuit (IC) containing all of the computation and control circuitry for a small computer. It provides economical computing power for many devices including home computers, TV remote control units, electronic tuners, cash registers and video games. Let's now look at the general operation of the microprocessor systems and associated building blocks.

DEVELOPMENT OF THE MICROPROCESSOR

The earliest electronic computers were built using thousands of vacuum tubes. These devices were very large and unreliable, and were mostly a laboratory curiosity. The next generation was built with transistors, which made computers much more reliable and reduced their size and cost. These "solid-state" machines marked the beginning of the computer as a practical device.

In the 1960s smaller, more powerful computers were built using hundreds of gates, flip-flops, and other similar integrated circuits. These ICs are called small scale integration (SSI) devices. As semiconductor technology developed, it became possible to put dozens of gates on a single IC. Examples of these medium scale integration (MSI) ICs are counters, decoders, registers, and adders.

This down sizing continues, and in 1971 the first microprocessor (the 4004) was introduced. Microprocessors contain the major computation and control sections of a computer, called the—Central Processing Unit (CPU), on a single integrated "chip." Microprocessors are often called microprocessing units (MPUs). A microprocessor chip contains thousands of gates and is called a large scale integration (LSI) device. LSI memory devices were also developed that store thousands of bits of digital information on a single IC. These two LSI devices made it possible to drastically reduce the total size and cost of small computers. Microprocessors have made it practical to build dedicated computers into many small, inexpensive products.

MICROPROCESSOR USE

Microprocessors are now being used in many products which were previously built with random logic devices. Microprocessor-based designs are usually less expensive and contain fewer components than the designs that they replace. Small microprocessor systems may be built with one or two ICs, at a cost of under ten dollars. These can often replace boards with dozens of simpler ICs. Because the number of discrete components and interconnections is greatly reduced, reliability is also improved.

This reduction in size is also possible using custom integrated circuits instead of microprocessors. However, the design of a custom IC can be an extremely complex and expensive process, often costing well over $100,000. This expense can be justified only for high volume products where the development costs can be spread over many thousands of units. The microprocessor allows standard ICs to be used to achieve the same miniaturization. The customizing takes the form of the program stored in the memory. Producing a standard memory with a custom program stored in it is a relatively inexpensive process.

The flexibility and power of microprocessor-based systems makes many sophisticated features possible, which in the past were impractical. For example, microprocessor-based systems can often test themselves to a considerable degree and give out error warnings. The microprocessor also makes practical the use of a keyboard instead of front panel switches. Another feature is the capability for complete remote control.

A BASIC MICROPROCESSOR SYSTEM

Think of a system with a keyboard and a numeric display, as in a pocket calculator. When a key is pressed, the corresponding number should appear on the display. This system is a natural application for a microprocessor, and in many ways similar to a minicomputer.

You will note in Fig. 8-1 a block diagram of a system for doing this. The microprocessor (also called the processor) is the "brains" of the system. It contains all of the logic to recognize and execute the list of instructions (program). The memory stores the program, and may also store data.

The microprocessor needs to exchange information with the keyboard and display. The input port, from which the processor can read data, connects the processor to the keyboard. The output port, to which the processor can send data, connects the processor to the display.

The blocks within the microcomputer are interconnected by three buses. A bus is a group of wires that connect the devices in the system in parallel. The microprocessor uses the address bus to select memory locations or input and output ports. You can

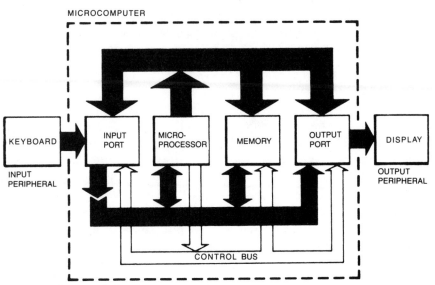

Fig. 8-1. Block diagram of microprocessor system.

think of the addresses as post office box numbers; they identify which locations to put information into or take information from.

Once the microprocessor selects a particular location via the address bus, it transfers the data via the data bus. Information can travel from the processor to the memory or an output port, or from an input port or memory to the processor. Note that the microprocessor is involved in all data transfers. Data usually does not go directly from one port to another, or from the memory to a port.

The third bus is called the control bus. It is a group of signals which are used by the microprocessor to notify memory and I/O devices that it is ready to perform a data transfer. Some signals in the control bus allow I/O or memory devices to make special requests from the processor.

A single digit or binary information (1 or 0) is called a bit (a contraction of binary digit). One digital signal (high or low) carries one bit of information. Microprocessors handle data not as individual bits, but as groups of bits called words. The most common microprocessors today use eight-bit words, which are called bytes. These microprocessors are called eight-bit processors. For an eight-bit processor, byte and word are often used interchangeably. Be aware, however, that word is also used to mean a group of sixteen or more bits.

Programs

A program is required by the system to perform the desired task. In the following example are some of the instructions required:

- Read data from the keyboard.

- Write data to the display.
- Repeat (go to step 1).

For the microprocessor to perform a task from a list of instructions, the instructions must be translated into a code that the microprocessor can understand. These codes are then stored in the systems memory. The microprocessor begins by reading the first coded instructions from the memory. The microprocessor decodes the meaning of the instructions and performs the indicated operation. The processor then reads the instruction from the next location in memory and performs the corresponding operation. This process is repeated, one memory location after another.

Certain instructions cause the microprocessor to jump out of sequence to another memory location for the next instruction. The program can therefore direct the microprocessor to return to a previous instruction in the program, creating a loop which is repeatedly executed. This enables operations which must be repeated many times to be performed by a relatively short program.

Peripherals

A complete microprocessor system, including the microprocessor, memory, and input and output ports is called a microcomputer. The devices connected to the input and output ports (the keyboard and display for example) are called peripherals, or input/output (I/O) devices. The peripherals are the systems interface with the user. They may also connect the microcomputer to other equipment. Storage devices such as tape or disc drives are also referred to as peripherals.

THREE-STATE DRIVERS

All devices in the microprocessor system exchange information with the microprocessor over the same set of wires (the data bus). The microprocessor selects one device to place data on the data bus and disconnect the others. It is the three-state output capability of the devices on the bus that enables the processor to selectively turn devices on and off.

In Fig. 8-2 you will see the symbol and truth table for a three-state buffer (often called a three-state driver). The buffer has an output enable in addition to the usual input and output. When the enable is low the buffer acts just as an ordinary buffer. The signal at the input is transferred to the output. When the enable is high, on the other hand, the output of the device is essentially disconnected.

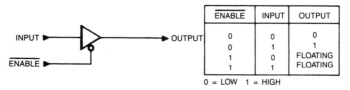

ENABLE	INPUT	OUTPUT
0	0	0
0	1	1
1	0	FLOATING
1	1	FLOATING

0 = LOW   1 = HIGH

Fig. 8-2. Three-state driver and truth table.

277

Looking at Fig. 8-3, you will see a conceptual equivalent circuit which generates the open state using a relay. The disabled (open) output state is often called the high impedance state. Refer to Fig. 8-4 for the schematic of a typical three-state output.

Three-state drivers are important because they allow many devices to share a single data line. This circuit allows any one of three different signals to drive one output.

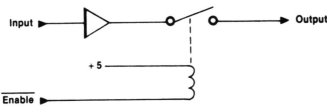

Fig. 8-3. Equivalent three-state driver.

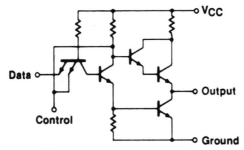

Fig. 8-4. Diagram of three-state driver.

Only one driver's enable line may be low, and that device drives the output. If more than one driver were enabled, they would both try to drive the output. This condition is not allowed because the logic state of the output would be unpredictable.

Many devices, including microprocessors and memories, contain internal three-state drivers. These ICs have an output enable, often called chip select (CS) or chip enable (CE), which controls their output drivers.

All devices which put data on the data bus have three-state drivers on their outputs. The microprocessor generates control signals (part of the control bus) to enable the three-state drivers of the device from which it wants to read data. The three-state drivers of the other devices are disabled.

THE MICROPROCESSOR

Illustrated in Fig. 8-5 are the basic signals that connect to a typical microprocessor. There are sixteen address outputs which drive the output bus, and eight data pins which connect to the data bus. The data pins are bidirectional, which means that data may go into or out of them. Read and write are the control signals that coordinate the movement of data on the data bus.

The two signals shown on the left of Fig. 8-5 provide additional control functions. The reset input is used to initialize the microprocessor's internal circuitry. The interrupt

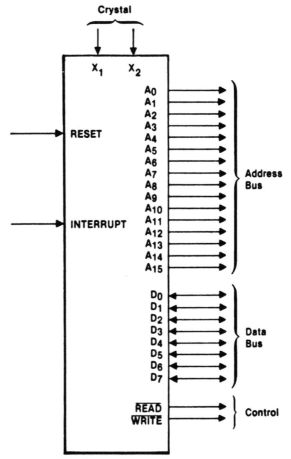

Fig. 8-5. Basic microprocessor signals.

input allows the microprocessor to be diverted from its current task to another task which must be performed immediately. The use of these signals, plus several others are also used.

The two connections at the top are for an external crystal, which is used to set the frequency of the oscillator in the microprocessor. The output of the oscillator is called the system clock. The clock synchronizes all devices in the system and sets the rate at which instructions are executed.

MEMORIES

Microprocessor systems usually use integrated circuit memories to store programs and data. They can store many bits of data in a single IC. Currently, devices are available with capacities well over 65,000 bits on one chip. A 65 k bit memory can store over eight thousand alphanumeric characters, or about three test pages on a piece of silicon about a third of an inch square.

The simplest memory device is the flip-flop, which stores one bit of information. Registers contain up to eight flip-flops on a single IC, each with its own data in and data out pins but with a common clock line.

LSI technology made it possible to put thousands of flip-flops on an IC, but a new problem was created. With thousands of flip-flops on an IC, there cannot be a separate data pin for each. The solution to this problem is to use address inputs to select the particular memory location (flip-flop) of interest. A decoder on the memory chip decodes the address and connects the selected memory location to the data pins.

Figure 8-6 shows a conceptual diagram of an eight-bit memory (most memories are much larger). Only the data output circuits are shown for simplicity. The decoder converts the binary address inputs to eight separate outputs, one for each possible combination of the three address lines. These signals control the three-state drivers at the output of each memory cell (flip-flop). The data from the addressed cell is placed on the data output line. This technique allows a single data pin to be used for all locations on the memory chip.

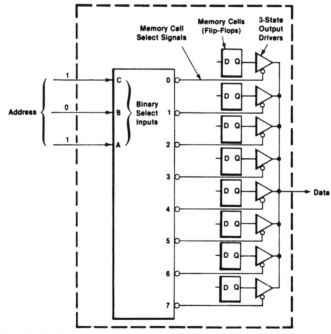

Fig. 8-6. Drawing of eight-bit memory.

Each memory location can contain a group of bits rather than just one bit as in the example above. Each can hold one, or eight bits, depending upon the particular IC. If the IC has eight data pins, then each memory location stores eight bits of data. Note that while the memory may contain thousands of locations, only one may be accessed at a time.

The number of addressable locations depends upon the number of address lines. With one address line, two locations can be selected: address 0 and address 1. With two address lines, one of four locations can be selected: 00, 01, 10 and 11.

The memory ICs used with microprocessors fall into two broad categories: ROMs and RAMs. A ROM (read only memory) is a memory which can only be read. The data is programmed into it at the time of manufacture, or by a special programming procedure prior to installation in the circuit. A program recorded into a ROM is often referred to as firmware.

A RAM (random access memory) is a memory into which data can be stored and then retrieved. RAM is actually a misnomer; random access means that the time to access any memory location is the same, a characteristic also present in ROMS. Read/write (R/W) memory is a more accurate term for what are usually called RAMs, but RAM is widely used to mean integrated circuit read/write memory. A digital tape recorder is an example of a memory that is not random access, since the time to access a particular location depends upon the position of the tape.

An important characteristic of semiconductor RAMs is that they are volatile: they lose their data when power is turned off, and when turned back on, they contain unknown data. ROMs do not have this problem, so they are used for permanent program data storage. Since the contents of a ROM cannot be modified, RAMs must be used for temporary program and data storage.

Figure 8-7 shows a ROM containing 2,048 words of eight bits each, or 16,384 bits. When using large numbers that are powers of two, "k" is often used to mean 1,024. Thus, this memory has 2 K bytes of 16K bits. Since each location contains eight bits, it is called a 2 k $\times$ 8 ROM.

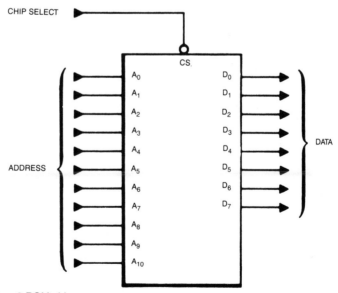

Fig. 8-7. A 2k $\times$ 8 ROM chip.

When the chip select input is low, the ROMs output drivers are enabled. When chip select is high, the data outputs are in the high impedance state. The three-state outputs allow the data lines of many memory devices to be connected together, with one device selected by bringing its chip select input low.

Illustrated in Fig. 8-8 is a 1 k × 8 RAM. This RAM contains 1,024 locations of eight bits each. The data lines are bidirectional, since data can go into or out of the memory. RAMs have an additional control line called write. To store data in the RAM, an address is selected, the data is placed on the data lines, and the write line is brought low. When the data and address are all set, the chip select is pulsed, and the data is stored in the memory. ROMs and RAMs come in many different sizes (with different numbers of words and different numbers of bits per word) and many types.

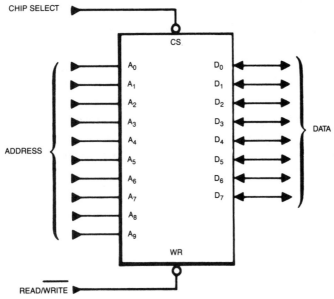

Fig. 8-8. Drawing of a 1k × 8 RAM chip.

MICROCOMPUTERS AND MINICOMPUTERS

A microcomputer is functionally similar to a minicomputer, and, in fact, the distinction between the two is becoming less clear each day. A microcomputer's CPU (central processing unit) is the microprocessor. A minicomputer's CPU is usually a PC board with dozens of less complex, but faster, integrated circuits on it. The main functional difference is that minicomputers are usually faster. They also are larger and more expensive. As microprocessors have increased in speed and power to compete with the older minicomputers, new minicomputers have been developed which are even faster and more powerful. These new minicomputers often use microprocessors internally. Thus, while the basic distinctions of speed, power, and size remain, the exact boundary is becoming vague. Microcomputers are now finding applications in systems where a minicomputer would be too bulky and expensive.

RADIO SHACK'S TRS-80 MICROCOMPUTER SYSTEM

In this section we will take a look at Radio Shack's very popular TRS-80 microcomputer system. Some brief system operation information will be given to help familiarize you with the system.

Troubleshooting information and tips will be given for the RAM, ROM, CPU, video divider chain, system clock, power supply, address decoder, address lines, keyboard and video processing system.

Also included in this chapter are the complete diagrams for the TRS-80 and some section isolation troubleshooting flowcharts.

Let's now take a brief look at the basic operation of Radio Shack's model TRS-80 microcomputer. Some basic theory of operation along with block diagrams will be covered. Some troubleshooting tips and system flowcharts will also be investigated, and some power supply chart information. Plus hints, ideas, suggestions and other tips. A photo of the Tandy 1000 microcomputer system is shown in Fig. 8-9.

Fig. 8-9. Tandy 1000 computer system. (Courtesy of Tandy Corp.)

TRS-80 System Block Diagram

You will find the integrated circuits contained in the TRS-80 can be broken down into ten major sections. The system block diagram in Fig. 8-10 shows these units as they are related to the other sections. The heart of the system is the CPU (central processing unit). Most of the leads on the CPU are data lines and address lines. The CPU tells the address bus where the data it wants is located, and the data bus is a good place for the information to come back to the CPU. The address lines are outputs from the CPU. They never receive data or addresses from other sections. The data lines on the other hand can give or receive data.

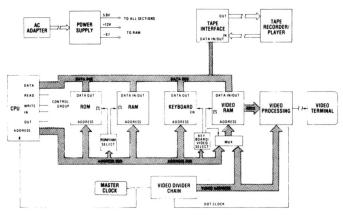

Fig. 8-10. System block diagram of Tandy TRS-80.

ROM Operation

If the CPU has to be the heart of the system, the ROM (read only memory) could very well be considered the brains. The ROM tells the CPU what to do, how to do it and where to put the data after each operation. Without the ROM, the CPU would just function or run as a pulse generator. When power is first applied to the system, the CPU outputs an address to the ROM that locates the CPU's first instruction. The ROM fires back the first instruction and then the two start communicating. In less than a second, the CPU, under ROM control, performs all the housekeeping required to start up the system and then flashes a "ready" on the screen.

If the CPU misses that first piece of ROM data, then it may become all fouled up. It may tell the ROM that it is ready to load a tape so that the ROM tells it how to do that. The tape recorder turns on. But since the CPU is now playing games in the video memory, the tape is not needed. Because the CPU operates at about 2 MHz these digital foul-ups seem to occur instantaneously.

You can just think of the CPU as the work horse and the ROM as the boss. The ROM tells the CPU how to do it, when to do it, and where it was placed.

RAM Operation

The next major section, see Fig. 8-11, is the RAM (random access memory). This memory is where the CPU can file data it may not need until later. The RAM is also the place where the programs are kept. If you tell the computer to count to 7,500, then the CPU stores your instructions in the RAM. The following occurs if you want the computer to act now.

The CPU tells the ROM someone wants in. The ROM tells the CPU to go to the keyboard and find out who. The CPU finds out, tells the ROM it's the number one chief. The ROM tells the CPU to find out what the chief wants. The CPU tells the ROM that the chief wants us to run. The ROM tells the CPU to go to the RAM and find out what the chief wants done. The CPU says the chief wants to count to 7,500. The ROM tells the CPU how to do it. After it's done, the ROM tells the CPU to find out what to do with it. The CPU informs the ROM that the 7,500 count must go on the video display screen and be saved. The ROM tells the CPU how to put it on the display and then tells it to store the 7,500 somewhere in RAM, then remember where it is. The CPU tells the ROM that the job is completed. The ROM tells the CPU to monitor the keyboard in case the chief wants another program run.

The CPU looks to the ROM for instructions. The CPU then follows the ROM's instructions and looks to the keyboard, then the RAM. In all cases, the CPU applies address locations to the ROM, RAM and keyboard. The data lines are then checked for input data that corresponds to these address locations. In case of an output from the CPU to RAM, the CPU selects the address, puts data on the data lines, and then instructs the RAM to store the data that is on the data lines.

Make a note, that only the CPU communicates with all other sections of the computer. If the CPU is told by ROM to store an item from ROM into RAM, the CPU can't make the RAM receive ROM data directly. Instead, the CPU takes the data from ROM and then sends it to the RAM. The CPU must act as intermediary between the two. The reason for this is that the CPU is the only section that can address locations and pass data to all other sections.

Keyboard and Video Processing

The keyboard section is not necessary as far as the CPU is concerned, but it is very necessary for data to be fed in by the operator. The keyboard is used for making known your instructions to the CPU. The opposite is true for the video RAM. In this case, the CPU wants to tell us it needs data or it may want to show us the result of a complex calculation. So, the request for more information or the result is stuffed into the video RAM. Anything in video RAM is automatically displayed on the terminal screen. The video processing section takes care of this. Data in the video RAM is in ASCII. Converting ASCII into the alphanumeric symbols we recognize is the task of the video processor. A ROM contains all of the dot patterns. The ASCII locates the character pattern, and the video processor sends it out to the terminal screen.

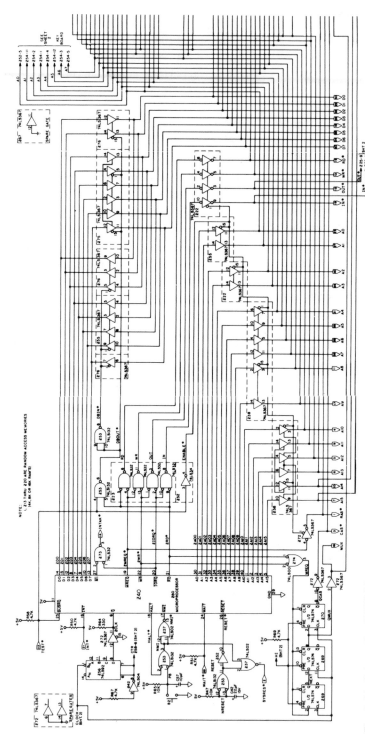

Fig. 8-11. Schematic of TRS-80 computer.

286

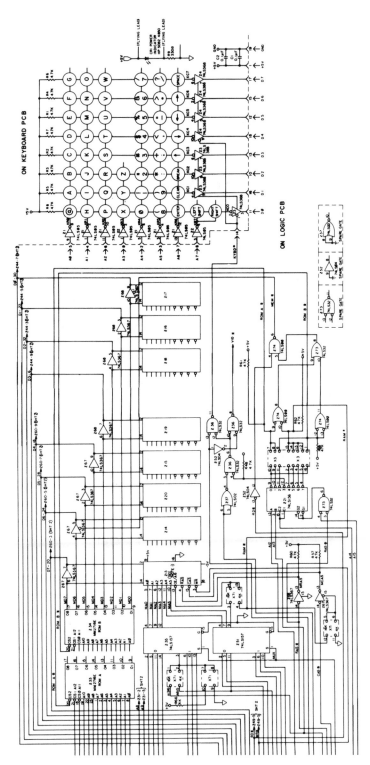

Fig. 8-11. Continued.

287

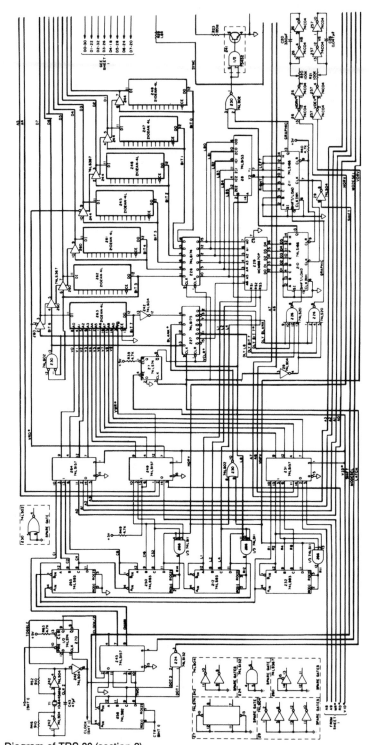

Fig. 8-12. Diagram of TRS-80 (section 2).

288

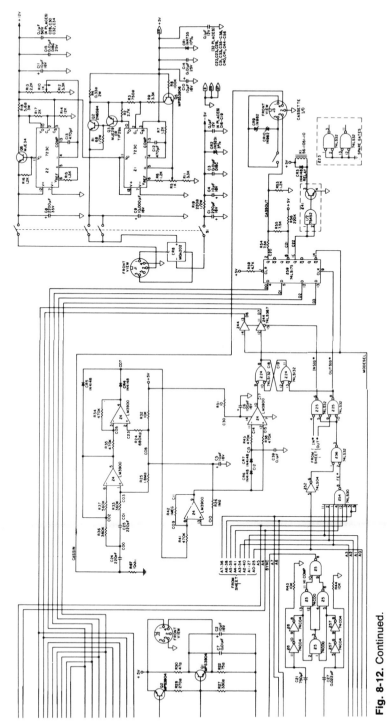

Fig. 8-12. Continued.

Video Divider Chain

Composite video going to a video terminal is extremely complex. Aside from the video signal, there is the horizontal and vertical sync. These signals must be very stable and be outputed in the correct sequence. The CPU is very busy, so the video divider chain handles the TV signal to the monitor. It generates the sync signals and addresses the video RAM in a logical order so that the video processor can handle video data efficiently. Note the block under the video RAM labeled MUX. This is short for multiplexer. It acts somewhat like a multipole, multiposition switch. When the video divider chain is in control, the MUX is switched so that only addresses from the divider chain are directed to the video RAM. The CPU may need to read or write data into the video RAM. If so, the MUX is switched so that the CPU has control over the video RAM's address. After the CPU is finished, the addressing task is reassigned to the divider chain.

System Clock

The system clock circuit will be found in Fig. 8-12. Y1 is a fundamental-cut 10.6445 MHz crystal. It is in a series resonant circuit consisting of two inverters. Z42, pins 1 and 2, and 3 and 4, form two inverting amplifiers. Feedback between the inverters is supplied by C43, a 47 pf capacitor. R46 and R52 force the inverters used in the oscillator to operate in their linear region.

The waveform at pin 5 of Z42 will resemble a sine wave at 10.6445 MHz. The oscillator should not be measured at this point, however, due to the loading effects test equipment would have at this node. Z42, pin 6, is the output of the oscillator buffer. Clock measurements may be made at this point. The output of the buffer is applied to three main sections: The CPU timing circuit, the video chain, and the video processing circuit.

Address Decoder

The memory mapped chart for the TRS-80 is shown in Fig. 8-13. Note that the address 01AC (in HEX) is in the ROM part of the map. Address 380A is in the keyboard area and 3CAA accesses the video display RAMs. Since the data and address buses are connected in parallel to all the sections, there must be some method to determine which section is being accessed. A decoding network monitors the higher order address bits and selects which "memory" the CPU wants to use.

The address decoder (Fig. 8-14) is very important to the operation of the system. Refer to this schematic since there are signals that need to be sourced or traced.

The address decoder uses six bits, A10 through A15 are needed, plus RD and RAS (row address select). A15 is the most significant bit of the address bus. Let's combine the six high order bits and add a couple more, so that we have two hex digits: A15, A14, A13, A12, A11, A10, A9 and A8.

	A15	A14	A13	A12	A11	A10	A9	A8	
From: Hex 0000	0	0	0	0	0	0	0	0	Level I ROMs
To: Hex 0FFF	0	0	0	0	1	1	1	1	
From: Hex 3800	0	0	1	1	1	0	0	0	Keyboard
To: Hex 38FF	0	0	1	1	1	0	0	0	
From: Hex 3C00	0	0	1	1	1	1	0	0	Display RAMS
To: Hex 3FFF	0	0	1	1	1	1	1	1	
From: Hex 4000	0	1	0	0	0	0	0	0	4K RAM
To: Hex 4FFF	0	1	0	0	1	1	1	1	
From: Hex 4000	0	1	0	0	0	0	0	0	16K RAM
To: Hex 7FFF	0	1	1	1	1	1	1	1	

Fig. 8-13. Mapped memory of TRS-80.

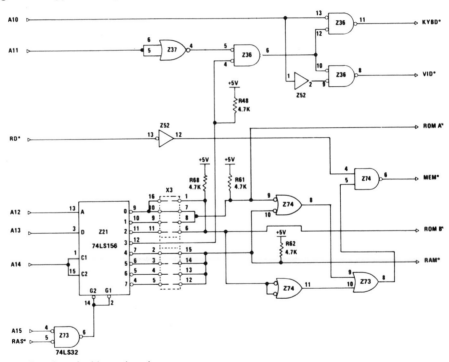

Fig. 8-14. Drawing of address decoder.

A12 through A15 form the most significant hex character. A8 through A9 are the two bits we had to add to complete that last hex character. Now let's break down part of the memory map into hex and binary. See Fig. 18-13.

Notice in the break down that we could use the two most significant digits of the hex code in the decoding scheme and handle the selection of all the memories. In the binary columns, you can see that instead of using two hex digits, which is eight binary lines, we can ignore two bits and use only six binary lines. A dotted line separates the two unused bits from the six that will be used.

Now look at Fig. 8-14 and you will see that bits A12, A13 and A14 are connected to Z21, a dual, 2-input to 4-line decoder/demultiplexer. The C1 and C2 inputs are connected in such a way as to make Z21 into a 3-input to 8-line decoder. The G1 and G2 inputs connected to Z21 are chip enables. As shown, when these inputs are at logical 0, Z21 is active. When high, Z21 is disabled and none of its eight outputs are low. The G-enables are controlled by OR gate Z73, pins 4, 5, and 6. Pin 4 is tied to A15, the most significant bit of the address bus.

Referring back to Fig. 8-14, then A15 and RAS are low at the same time, a low will be outputed by Z73, pin 6. This low will enable Z21. When Z21 turns on, one of its outputs will go low, depending on the status of A12, A13, and A14. For example, if these three inputs are high, pin 4 will go low. If all three inputs are high, pin 4 will go low. You might consider A12 through A14 as supplying an octal address to Z21. Since there are eight states in an octal code, then there could be one of eight lines selected (output 0 through output 7).

We can sum up Z21's function quite simply. It decodes the most significant digit of the hex address. Using Z21 and the last two bits, A11 and A10, we can define any one of the four "memories" available to the CPU in level I.

Address Decoder Programming

Attached to the outputs of Z21 is X3. X3 is called a "dip shunt" and it is installed in the PCB position Z3. A dip shunt is like a shorting bar array, except the bars may be broken. By breaking some bars and leaving others intact, the address decoder is programmed to reflect the amount of RAM and ROM the CPU has available for use. You will note in Fig. 8-14, X3 is shown with six broken shorting bars.

Power Supply System

The TRS-80 requires three voltage supply levels as follows.

A + 12 volts at 350 milliamps.

Also, +5 volts at about 1.2 amps.

And a −5 volts at 1 milliamps.

The +12 and −5 volts are needed by the system's RAM and all other devices that require +5 volts. The +12 volt and +5 volt supplies are regulated and current-protected against shorts. The −5 volts supply is not as critical as the other two supplies, and it uses a single zener diode for regulation. The stepped-down ac voltage is supplied by an "ac adaptor." The adaptor has a transformer with one primary and two secondary windings.

The secondary windings are both center-tapped. One is rated at 14 volts ac at one amp. This winding is used for the +5 and −5 volt supplies. The other winding had diodes connected and it outputs 19.8 Vdc at about 350 milliamps. This circuit is used for the 12 volt supply. All voltage outputs and center taps are brought into the power input jack.

Unregulated dc voltage for the +12 volt supply is inputed at pin 2 of the power jack. When the power switch is closed the output voltage is about 20 volts due to action of the filter capacitor. This voltage is then fed to a transistor and regulator, Z2. Figure 8-15 shows a simplified diagram of the internal circuitry for the 723 regulator chip.

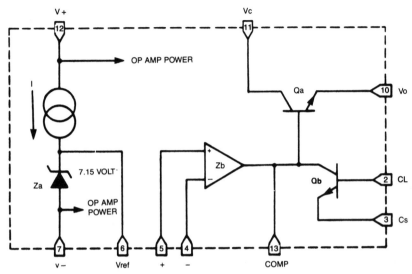

Fig. 8-15. Block diagram of 723 regulator Z2.

Referring to the circuit in Fig. 8-16, we see the filtered dc voltage from the adaptor and C8 is applied to pin 12 of Z2 and the emitter of series pass transistor, Q6. The voltage applied to pin 12 allows a constant current source to supply zener current for Za. Pin 6 of Z2 will output a zener voltage of about 7.15 volts. Pin 6 is tied to pin 5, the positive input to operate amplifier Zb. The negative input to the op amp is tied to the wiper of R10. Initially, pin 4 of Z2 is at ground, forcing the output of op amp Zb to output about 7.15 volts. Transistor Qa turns on, which turns on pass transistor Q6. The pass transistor supplies voltage for current monitoring resistor R18 and to the resistor network R13, R10 and R12. If R10 is adjusted for 7.15 volts at its wiper, the op amp will be balanced and Q6 will output only enough voltage to keep the loop stable. If the output voltage dropped below 12 volts, Zb's output would decrease which would force the current through Qa to decrease. Qa would cause Q6 to increase the current through it, and the output would rise back up to the 12 volt level. If the 12 volt line increased in voltage, the op amp would cause Qa's current to increase, forcing Q6 to drop down.

Power Supply Checks for the TRS-80

Most problems that result in loss of power supply voltage will be associated with solder shorts, component shorts or defective power supply adaptors. Usually, the

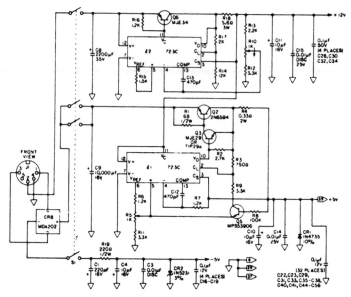

Fig. 8-16. Computer power supply circuit.

power supply will not be damaged due to a short because the regulators use current-limiting with fold-back. A solder short or shorted component does not have to be in the power section to cause a supply problem. The short may be anywhere.

Should you be missing +12 volts and +5 volts, measure the voltage across R18. This resistor monitors the current flow from the +12 volt supply. If the voltage reads about 0.6 volts, the +12 volt bus is in fold-back and has shut itself off. Since the +12 volt bus is shut off, you will not have +5 volts because the +5 volt regulator is referenced to the +12 volt output. You must now locate and remove the short on the +12 volt bus before anything will operate.

If the −5 volt supply is not present, first confirm that there is ample negative voltage on the adaptor side of R19. See if R19 is dropping all of the voltage. If so, you have a −5 volt bus short.

The +12 volt and the −5 volt supplies are used by the RAM system. If you have problems with either of these two, suspect a RAM short. See if you can find a RAM that generates more heat than the others. Use a temperature probe for this check.

Pull all RAM's and retest. If all of the power supplies are now good, turn off the power and reinstall one RAM. Turn on the power and retest. Install each RAM until you find one that loads down the power supply. Remove the faulty RAM and continue to check out the other IC's. There may be more than one shorted device.

A short on the +5 volt bus can be a real tough one to find. Unless you can see the short, you will have to cut PC runs to isolate a section. Once a section is isolated you will probably have to make other cuts to find the short. And, remember to bridge the cuts when repairs are completed. These runs carry considerable current, so use solid 22 gauge wire to bridge the cuts.

If you find a dead +12 volt bus, check transistor Q1's heat sink. The hardware holding the transistors heat sink may have become loose and shorted Q1's leads.

Table 8-1 lists the voltages found around Z1 and Z2 for a normal operating unit. These measurements were taken when the +12 volt supply was adjusted for 12 volts and the −5 volts supply was adjusted to 5 volts.

Table 8-1. Power Supply Voltage Chart.

Z1		Z2	
Pin Number	Voltage	Pin Number	Voltage
1	0.00	1	0.00
2	5.30	2	10.60
3	5.00	3	11.99
4	5.00	4	6.92
5	5.00	5	6.92
6	7.46	6	6.92
7	0.00	7	0.00
8	0.00	8	0.00
9	0.33	9	5.72
10	5.89	10	12.31
11	11.99	11	21.16
12	11.99	12	21.69
13	7.05	13	13.48
14	0.00	14	0.00

All voltages are measured with a digital voltmeter. Voltages are referenced to ground at the right side of capacitor C9.

Isolation of a Defective Computer Section

One of the toughest problems to section-isolate is a screen full of junk at POWER ON. This is a display with all character positions filled with either alphanumerics or graphics. Also, a garbage condition does not always indicate that the power-up logic is defective. A problem could exist in RAM, ROM, the video divider chain and of course the CPU itself. Thus, a fault could exist in 75% of the computer system.

You could start by replacing all of the RAM's, ROM's and the CPU but, this would be a waste of time. If the problem was a cold solder joint or a short, replacing all plug-in devices would not solve the problem. The section isolation technique will probably yield much more positive results. This is based on a removal technique that eliminates sections from the suspect list.

Section Isolation Flowchart

Referring to Fig. 8-17, you will find a flowchart of section isolation "by part removal". Start this process in the parallelogram, block 1, which gives the basic problem. Block 2 instructs you to disassemble the unit and reconnect the video and power inputs. Block 3 is a decision block. Do you have garbage on the screen now? If so, you continue to block 5. If not, block 4 tells you to suspect a shorting interconnect cable between the

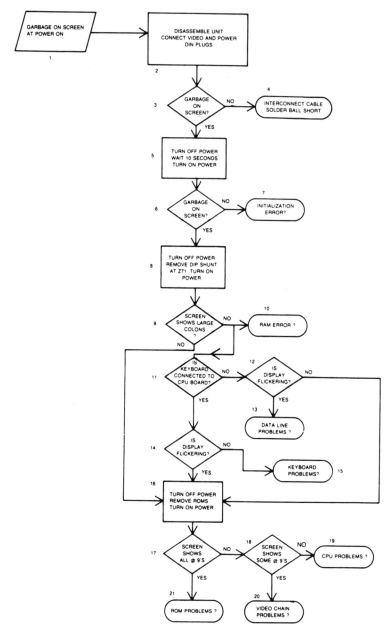

Fig. 8-17. Part removal isolation flowchart.

keyboard and the CPU board. You could also have loosened a "solder ball" during disassembly, and the short is now gone. Examine the interconnect cable carefully for shorting conductors. Also, note any loose solder or wire bits on the board. You may have solved the computer problem just by taking it apart. So, run another test.

At block 5, you will turn off power to the unit, wait about ten seconds, then switch power back on. The delay gives the initialization logic time to reset. If there is now a "ready" on the screen at block 6, you may have a problem around S2 or C42, as block 7 instructs.

In block 8, you are instructed to remove the dip shunt (X71) at Z71. Refer to fold out diagrams. With X71 removed, the RAM's are not electronically in the system. When power is applied, the ROM and the CPU are in communication, but there is no data flow to or from the RAM.

The screen should show a pattern of 16 character lines of 32 colons. If the CPU shows large colons, you could have RAM or keyboard-type problems. Blocks 11 through 15 will help in isolating that type of problem. As blocks 12 and 14 imply, there are two colon displays. One display is stable. The other is blinking and flickering as the CPU constantly interrupts video addressing. Depending on the status of the keyboard, you could have data line or keyboard troubles.

The next step at block 16 is to remove the ROM's. The CPU is now locked up without instructions from ROM. The pattern to look for is a screen full of @9's. The display should be in 64 character format at this time. The display will continually be flickering.

If you get @9's on the screen, you probably have a ROM error. If no @9's or partial @9's are visible, you could have video chain or video RAM problems. If you still get garbage, maybe the CPU is dead or something is making the CPU not function.

As we see, the part removal isolation technique uses a lot of maybe's, question marks and could be's. The "what if" are trying to tell us what section could be at fault. You could have ROM problems and yet obtain large colons. You could get @9's and still have CPU error. But it is better than nothing, and the process does give you a starting point.

Signal Condition (Active, Steady-State, or Floating)

Normal troubleshooting techniques call for an output-to-input sweep of the bad signal line. Hence, once a bad signal is found, the circuit is traced backwards until the signal is correct. The failed device will be located between the good input and the bad output.

Active. Defined as any logical transition from high to low or vice versa. For example, the output of oscillator buffer Z42, pin 6, always has activity. There is a constant output pulse train at this pin. The signal swings from almost ground to over 3 volts continuant.

Steady-State. Defined as a logical 1 or logical 0. For example, Z40, pin 16, has a steady-state logical 1. It is held high by resistor R50. Another example is the logical 0 at pins 6 and 7 of Z56, the CPU clock divider. Z42, pin 8, is always low unless resistor R67 is grounded.

Floating. Defined as a signal level between the steady-state of a logical 0 and a logical 1. The CPU, the ROM's, the RAM's and the data and address buffers are all tri-stable devices. When tri-state devices are disabled or unselected, the output may show a floating condition. In a floating condition, the output will show system noise flickering through it. The average level of the noise will attain a voltage of 1.5 volts or so. TTL devices define a logical 0 to be equal to, or less than, 0.8 volts. A logical 1 has a voltage equal to, or more than, 2.4 volts. Any voltage between these two levels will be considered floating.

CPU Problems

A problem with the CPU does not mean that Z40 is inoperative. It could indicate you have trouble with the address and data buffers, the control group, CAS/RAS timing, or with one of the CPU's support devices. If you suspect a problem with these devices, try substitution of a known good CPU for Z40. The flowchart, shown in Fig. 8-18 will help you for CPU troubleshooting.

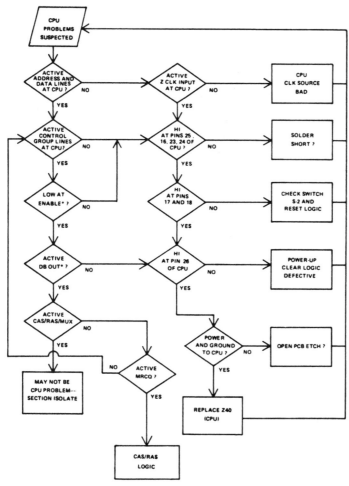

Fig. 8-18. CPU section isolation flowchart.

298

The primary objective of this chart is to help you find a signal fast that should be active but is not. The main flow of the chart is on the left side of the blocks. Here, you are checking for activity on address and data lines. With no activity on the address lines, you are immediately branched off to the CPU's support group to find out why. Pay particular attention to the appearance of address line outputs. Any tri-state looking signal could mean a potential short between address lines. The opposite is true of data lines. These signals may be active and have floating components between active states. Hence, data line shorts are extremely difficult to find, using an oscilloscope.

Addressing Problems (ROM)

Usually addressing problems are associated with open or shorted address lines going to the ROM sockets. Early versions of the boards may have jumper modifications on the solder side that may have broken loose. There is also the chance that vibration has jarred a ROM partially out of its socket. The address lines should be checked at the chip. Normally, there will be activity on all lines.

There are two types of data problems. The first one is the non-repairable bit error internal to the ROM. The checksum contained in the SCQATS program can readily verify this. SCQATS is a special machine language test and debugging tape available from Radio Shack for the TRS-80. If the ROM problem is too severe for SCQATS loading, a replacement test may be required. The second type of data problem is the short or open on the data output. If you remove dip shunt Z3, the ROM's will tri-state and you can check for a floating state on the data pins.

RAM Checks

RAM problems are slightly more difficult to troubleshoot because of multiplexing of the address inputs. Aside from addressing differences, the RAM's are checked like the ROM's. If you have a RAM problem and the system will not load SCQATS, you can replace the eight RAM's with a known good set. If this clears the faults, start replacing your standard RAM's with the parts you took out, one by one. Power up after each exchange to see if you still have a "ready". Continue this process until you have isolated the defective RAM(s).

Address Decoder Tips

A problem in the address decoder section will probably point you in the memory direction. For example, if the ROM is never addressed with ROM, you would think you have ROM problems. If you suspect one of the "memory" locations, keep in mind that the address decoder sources what the memory selects. The select inputs to the different memories should be the very first thing to check out.

Since the address decoder is made up of gates, it should be easy to repair once you locate the fault. The hard part is knowing when to suspect a fault with the decoder section. Section isolation demands that the address decoder be functional, at least partially. Unfortunately there is no "cut and try" way to determine if this section is operating correctly. Of course, you can monitor each output to see if it's responding, but you really cannot be sure the signal is supposed to be there when it is.

Keyboard Information

Difficulty with the keyboard is usually mechanical, sticking keycaps, bouncy keys and broken interconnect cables are common. Shorts in the keyboard matrix are usually easy to detect. If you find an alphanumeric character display right after the 7, that particular key, or PCB run, may be shorted. A completely "dead" keyboard could be caused by lack of power, a broken interconnect cable or the address decoder is not feeding a signal to the keyboard.

Video Divider Chain Comments

Problems in the video divider chain will usually be associated with the stability of the display. Loss of vertical or horizontal reference frequencies sometimes can be traced back to defective counters or bad reset gates. Since the systems master clock/oscillator is included in this section, inactive (dead) system troubleshooting can very well end up at this point.

Since most of the reference and timing signals for the video processor are generated in the divider chain, most (but not all) display difficulties can be isolated to this section. This is especially true of vertical roll or horizontal tear of the display. If the horizontal or vertical reference frequency is not getting to the sync processors, then the problem will definitely be a divider chain foul-up.

Notes on the Video RAM's

If you suspect video RAM problems, you should try a SCQATS tape test loading. SCQATS will be most helpful in rooting out bit-error in the RAM's. If the test generates large amounts of bit-errors, you should suspect either the divider chain or the video RAM addressing multiplexer.

Normally, addressing errors occur when there is a short or open between the multiplexer and the RAM's. Signal activity on the address inputs of the RAM's can be easily checked with an oscilloscope. All address lines (V0 through V9) should be active in 64 character format. There will not normally be any floating conditions on these inputs. The logic input to video RAM will only be active during a NPU data transfer. Normally, it should be high.

Video Processing

Problems in the video processing section can range anywhere from a blank screen to missing dots. Usually, the fault is easily found because this section is a serial-type. For example, if you have graphics problems, you know there are only two chips that are used as graphics handling devices. You should then look around shift register Z11 and graphics generator Z8. The parts that are strictly alphanumeric are character generator Z29 and its shift register Z10. Defective devices that can affect both alphanumeric and graphics are: Z26, Z27, Z30 and the video mixing circuits, consisting of Q1 and Q2.

Sync Generator Points of Service

The sync generator section is one of the easiest circuits to troubleshoot. If the timing reference is getting to Z6 and Z57, it is a simple process to find the point where you have lost the signal. A problem can occur with the adjust pots, R20 or R21. Severe heat buildup may cause these parts to fail. Capacitors C20 and C26 are usually dependable unless they are physically damaged. You may find C21 or C27 shorted. These capacitors are mylar and are very susceptible to shorting out under impact stress.

An important point about this circuit is that Z6 and Z57 are CMOS devices. Unlike TTL, they are high impedance devices that consume little current. A floating condition on a CMOS input will not necessarily give a floating "display" on an oscilloscope. A floating condition may look high or low depending on the charge of the broken line tied to the input point. Even the resistance of your finger across a broken run can complete the circuit and cause a CMOS device to operate. When you remove your finger from the run or pin, circuit operation may fade away very slowly as the PC board run discharges.

PROCESSOR TECHNOLOGY CORP. SOL-20 MICROCOMPUTER

Some of the special circuits found in the Processor Technology SOL-20 Microcomputer shown in Fig. 8-19 will now be covered.

Fig. 8-19. SOL-20 is the first computer built by the author.

Power Supply Circuit

The fused primary power is applied through S5 to T2. Now refer to Fig. 8-20 for a block diagram of the SOL-20 power supply. A full-wave bridge rectifier, is connected across the 8-volt secondary of the transformer. The rectified output is filtered by C8 and applied to the collector of Q1. Q1, a pass transistor, is driven by Q2, with the two connected as a Darlington pair. The output of Q1 is connected to R1 which serves as an overload current.

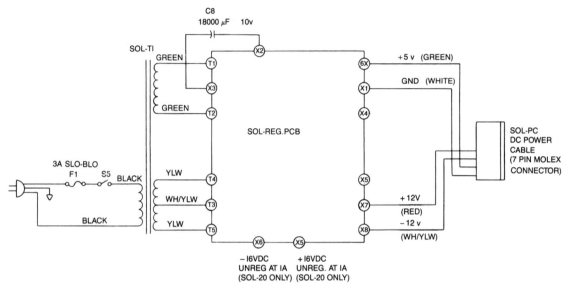

Fig. 8-20. SOL-20 power supply block diagram.

For the operation of the SOL-20 power supply refer to circuit in Fig. 8-21. An overload current (approximately 4 amps) increases the voltage drop across R1. The difference is amplified in one-half of U2 (an operational amplifier) and the output on pin 7 turns Q3 on. Q3 in turn "steals" current from Q1-Q2 and diverts current from the output on pin 1 of U2. This in effect turns the supply off to reduce the current and voltage. Note that the circuit is not a constant current regulator since the current is "folded back" by R6 and R8. The current is reduced to about 1 amp as the output voltage falls to zero.

Divider network R11 and R12, which is returned to −12 volts, senses changes in the output voltage. If the output voltage is 5 volts, the input to pin 2 of U2 is at zero volts. U2 provides a positive output on pin 1 if pin 3 is more positive than pin 2 and a negative output for the opposite condition.

When the output voltage falls below 5 volts, pin 2 of U2 goes more negative then pin 3. This means pin 1 of U2 goes positive to supply more current to the base of Q1. The resulting increase in current to the load causes the output voltage to rise until it stabilizes at 5 volts. Should the output voltage rise above 5 volts, the circuit operates in a reverse manner to lower the voltage.

302

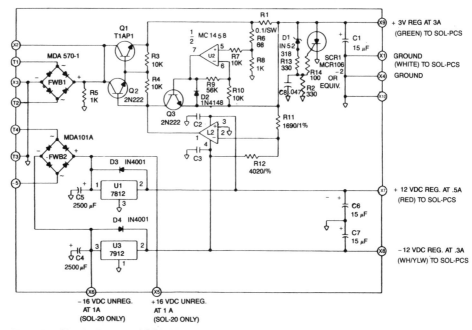

Fig. 8-21. Circuit diagram of SOL-20 power supply.

Protection against a serious over-voltage condition (more than 6 volts) is also provided. Zener diode, (D1), with a 5.1 Zener voltage, is connected in series with R13 and R2. When the output voltage exceeds about 6 volts, the resulting voltage drop across R2 triggers SCR1 to short the fold-back current to ground. Since the overload current circuit is also working, the current through SCR1 is about 1 amp. Once the current is removed, this circuit restores itself to its normal condition; that is, SCR1 turns off.

Bridge rectifier FWB2, connected across the other T1 secondary, supplies +12 and −12 volts dc. The positive output of FWB2 is filtered by C5 and regulated by IC regulator U1. The negative output is filtered by C4 and regulated by U3. Shunt diodes D3 and D4 protect U1 and U3 against discharge of C6 and C7 when power is turned off. (Note that should the −12 volt supply short to ground, the +5 volt supply turns off by action of U2).

Unregulated −16 and +16 volt dc, from the filtered outputs of FWB2 are available on terminals X6 and X5. Note the power transformer (T2) has an 8 volt secondary winding and a bridge rectifier (FWB3) to supply +8 volts dc at 8 amps. This output is filtered by C9 and controlled by bleeder resistor R13. The SOL-20 also has a cooling fan powered by ac line voltage.

SIGNATURE ANALYSIS

Signature analysis is a troubleshooting technique that was introduced by Hewlett-Packard several years ago. It is based on the principle that a good digital circuit in a given state will produce the same output when stimulated repeatedly by the same input. Should the repeated output of a device not be the one it was designed to produce, then it is defective.

While this technique is simple and very obvious, signature analysis is more complex, and uses mathematic redundancy coding. When a micro system has been designed with signature analysis in mind then troubleshooting can be quite simple.

All a technician need do on the bench or in the field, is follow the old technique of signal tracing. You just check one test point and compare the measurement result to a table or schematic. If it does not match, you check another point as indicated in the test plan.

The point between the last bad measurement and the next good one is the fault location. Note that you do not need to know anything about signature analysis or even digital logic.

In the past, a repetitive digital stimulus, or bit stream, was generated by the microprocessor of the system under test, which would start a special test program built into the on-board memory. A signature analyzer is now used to check out the response of the board's various test points.

The analyzer works by monitoring the bit stream on one line for a certain period determined by the clock rate of the circuit under test. In this process it compresses the data and translates it into a four-character hexadecimal word, or signature.

A set of signatures must be generated, one for each test node in the circuit. The test engineer can do this by operating the unit with routines that exercise each area of the board. The good signatures are recorded, and the board is analyzed for faults that affect the bit stream. When this is completed the design engineer can create test procedures and diagram test points to be used by field technicians.

The three key conditions for the application of signature analysis are as follows:

- Initialize circuits that can hold two or more different states (RAM's, flip-flops, and counters for example).
- Synthesize the timing needed for signature sampling, both in terms of framing the sequence (sample start and sample stop) and clocking the bit stream.
- Application of the stimulus.

HEWLETT-PACKARD 5004A SIGNATURE ANALYZER

The HP 5004A signature analyzer checks out a compatibly designed digital product by detecting the bit streams at various circuit nodes and displaying them as hexadecimal signatures, which may or may not agree with the correct values noted on the schematic. It is a lightweight, portable instrument, built around the feedback-shift-register circuitry that produces the signatures, and it is equipped with an active probe for data input.

The probe has dual threshold levels that are compatible with transistor-transistor logic. It also acts as a TTL probe and in this way provides additional troubleshooting information by indicating high, low, mid-level, and pulsing states.

Via an active "pod" on the 5004A's input cable, the unit test supplies the instrument with three gating signals: start, stop, and clock. Start signals the beginning of a measurement window, preparing the shift register to receive information from the data probe. Stop closes the measurement window. Clock is the system clock of the product under test and assures synchronous acceptance of data and gating signals by the shift register. The active edge of each of these gating signals can be selected at the front panel, giving the system more latitude.

The front panel also includes a gate light and an unstable signature light. The gate light indicates proper start/stop gating operation, remaining on during the measurement window, with stretching to make it visible to the operator during very short ON times. The unstable-signature light indicates a difference between signatures in adjacent windows, alerting the operator to intermittent problems that may not be apparent from the display.

Two other controls are the hold and reset switches. The hold feature allows observation of single-event (one-shot) signatures, such as a power-on-restart routine. The 5004A will display only the signature associated with the first valid window and will hold the display until the probe reset switch is pressed. The hold/reset controls are also useful for taking signatures in difficult locations where you cannot watch the probe and display at the same time.

Because the 5004A is synchronized to the system clock of the device under test (up to 10 megahertz), setup time becomes important. Setup time is the period during which data must be stable before arrival of the selected clock edge. In the H-P-5004A, the maximum data setup time is 15 nanoseconds but usually runs about 8 ns. This leaves the balance of the clock cycle for the settling times of components in the device under test. No hold time is required after the selected clock edge.

Any data state changes between clock edges are disregarded. The first data bit accepted during a measurement window is the one coinciding with the synchronized start-signal edge. The last bit accepted is that preceding the synchronized stop-signal edge as shown in the Fig. 8-22 waveform drawing.

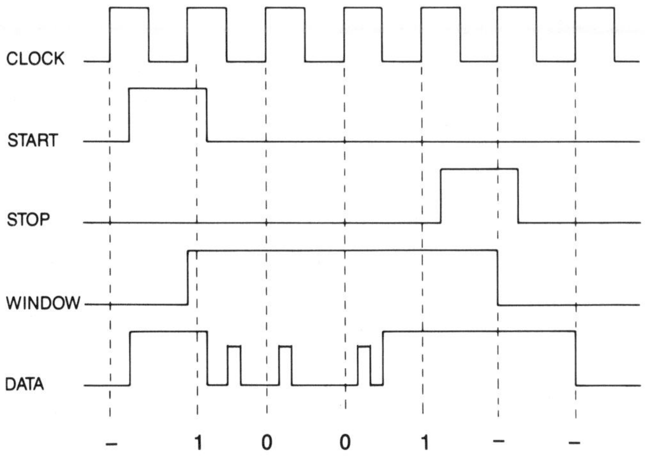

Fig. 8-22. Waveforms show the laser bit accepted is that preceding the synchronized stop-signal edge.

Tri-state bus architectures are common in many types of equipment and pose the problem of how to interpret the floating state for the purposes of consistent signature detection. Pullup resistors in the circuit under test would force a bus high in the third state, but would slow down the state transitions and possibly cause inconsistent signatures. Instead, the 5004A data probe pulls to the 1.4-volt level through a 50-kilohm input resistor and employs hysteresis. This causes the floating state of a tri-state bus to be entered without ambiguity into the signature as the same state (0 or 1) as the most recent valid bit.

To increase the confidence of on-site service, the front panel self-test feature allows a go/no-go checkout of the entire 5004A, including probe, pod, and cables. An internal program work out the analyzer and displays the results. If this display indicates a malfunction, the 5004A itself can be switched to its own test mode and diagnosed to the component level with another signature analyzer.

The Signature Analyzer Technique

Let's now look at a H-P instrument that uses signature analysis, this is the H-P 3455A voltmeter, which has an extensive digital portion. It is microprocessor controlled and contains a self-test program stored in ROM. If the self-test fails, a jumper inside the enclosure is removed, breaking feedback loops and also enabling the signature-analysis routine that is used to diagnose the instrument.

The decision to go with signature analysis made the design easier to troubleshoot down to the component level. All of the digital portion is on one board. The elimination of connectors and a batch of smaller PC subassemblies reduced the production cost and also made all the parts easier to test without special extender boards.

Some extra design time, a few more ROM locations, and the extra jumper wire were the price paid for this kind of serviceability. The cost evaluation proved that the production costs actually were lower.

Signature Analyzer Service Manual

The service manual is written so that a technician unfamiliar with the signature analyzer can walk up to a faulty voltmeter, read the instructions, and within a short time locate the problem. One element in the manual is a troubleshooting flowchart or tree shown in Fig. 8-23, which systematically guides the technician through the fault finding process.

The initial tests rely very little on signature analysis, yet may allow isolation of the fault down to a specific area. Diagnostic programs cannot carry on from here, since they do not have access to individual nodes, but it is from here on down to the components that signature analysis excels.

At this level, the technician uses the schematics and graphs of board layouts together with the flowchart, to find the defective node, in some cases the voltmeter (the first test checks its kernel which consists of the microprocessor, the clock circuit, the power supply, and a number of external gates). After the proper functioning of the

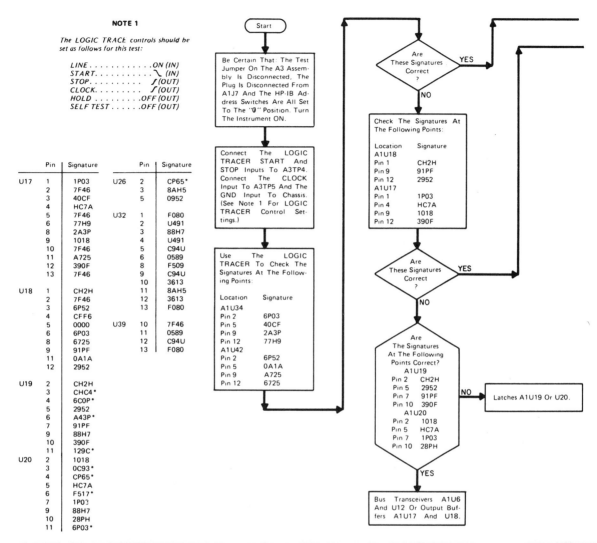

Fig. 8-23. This TROUBLESHOOTING TREE for the first instrument to use signature analysis contains a list of signatures that should be at the designated pin numbers of the various devices. (Courtesy of Hewlett-Packard)

kernal has been verified, the test setup is changed (one control wire of the logic tracer is moved to another pin in the 3455A), and the remaining portions of the circuit are tested. A special portion of the ROM control loads and reads the random-access memories. Some asynchronous portions require a third test setup. Again, the connection of the start is just moved to the next pin designated for this purpose, and troubleshooting can be continued.

For documentation the service manual shows pictures of the board and defines the setup for each test. Each picture shows only the signatures related to the particular test, directing the technicians checks to the important areas on the board. The ROM program even simulates interrupt signals, ensuring that they occur predictably at the same spot within a window so that stable signatures will result.

The signature analyzer has its own self-check. Each test setup is tested by touching the power supply voltage with the instrument's probe to input a sequence of all 1s. If this characteristic signature is correct, the setup conditions and the framing of the measurement window are verified. Specifically, this tells the user that the switches on the signature analyzer, as well as all the jumpers, switches, and control buttons in the front and rear of the voltmeter are set correctly. Thus, the confidence level of the technician is very high at the start of a test.

This first application of the H-P signature analyzer went very smoothly, because the engineer developing service techniques worked closely with the design engineer. Generally, the design engineer wrote the stimulus routines, while the service engineer involved himself with documentation and overall test strategy. This early involvement also ensured that the designer, with the many demands on his time, did not neglect to think about serviceability and, as an example, put off allocating read-only-memory space, perhaps too late in the design cycle.

The fact that signature analysis is built into the voltmeter also made final testing on the production line much easier. At H-P the signature analyzer is now a favorite piece of production line test equipment. Thus, the signature analyzer will have a significant impact on present and future service costs.

Index

313